Circuit Modeling of Inductively-Coupled Pulsed Accelerators

This monograph describes lumped-element modeling techniques for inductively-coupled pulsed accelerators, starting from the basic physical description of the various processes and then bringing all the pieces together into solutions. Coilguns, inductive pulsed plasma thrusters, and compact toroids have each been individually studied using the methods used in this monograph.

This monograph is of interest to researchers and graduate students in physics, engineering, and mathematics presently studying inductively-coupled pulsed accelerators.

Circuit Modeling of Inductively-Coupled Pulsed Accelerators

Kurt A. Polzin
Ashley K. Hallock
Kamesh Sankaran
Justin M. Little

CRC Press
Taylor & Francis Group
Boca Raton London New York

CRC Press is an imprint of the
Taylor & Francis Group, an **informa** business

First edition published 2022
by CRC Press
4 Park Square, Milton Park, Abingdon, Oxon, OX14 4RN

and by CRC Press
6000 Broken Sound Parkway NW, Suite 300, Boca Raton, FL 33487-2742

CRC Press is an imprint of Informa UK Limited

British Library Cataloguing-in-Publication Data
A catalogue record for this book is available from the British Library

ISBN: 978-0-367-34900-4 (hbk)
ISBN: 978-1-032-37410-9 (pbk)
ISBN: 978-0-429-35197-6 (ebk)

DOI: 10.1201/9780429351976

Publisher's note: This book has been prepared from camera-ready copy provided by
the authors.

Typeset in Computer Modern
by KnowledgeWorks Global Ltd.

Contents

■ Contents

Author Biographies

Kurt A. Polzin is Chief Engineer for Space Nuclear Propulsion at NASA's Marshall Space Flight Center. He received his Ph.D. in Mechanical and Aerospace Engineering from Princeton University in 2006. Dr. Polzin specializes in electric thrusters and space nuclear power and propulsion systems. He is a Senior Member of the Institute for Electrical and Electronics Engineers (IEEE) and an Associate Fellow of the American Institute of Aeronautics and Astronautics (AIAA).

Ashley K. Hallock is the Electric Propulsion Subsystems Lead Engineer at OHB Sweden, working with teams to develop electric and chemical propulsion subsystems for a wide range of satellites. She received her Ph.D. in Mechanical and Aerospace Engineering from Princeton University in 2012. Dr. Hallock is a Senior Member of the American Institute of Aeronautics and Astronautics (AIAA) and serves on the Nuclear & Future Flight Propulsion and Plasmadynamics & Lasers Technical Committees.

Kamesh Sankaran is a Professor of Engineering & Physics at Whitworth University. He received his Ph.D. in Mechanical and Aerospace Engineering from Princeton University in 2005. Prof. Sankaran has been on the faculty at Whitworth since 2004, where he specializes in spacecraft propulsion, plasma physics, computational physics, shock physics, and public policy in science and technology. He is a Senior Member of the American Institute of Aeronautics and Astronautics (AIAA).

Justin M. Little is an Assistant Professor of Aeronautics & Astronautics at the University of Washington. He received his Ph.D. in Mechanical and Aerospace Engineering from Princeton University in 2014. His research interests focus on novel electric propulsion concepts with applications to space exploration, telecommunications, and low-cost small satellites. Prof. Little is a recipient of an AFOSR Young Investigator Award and serves on the American Institute of Aeronautics and Astronautics (AIAA) Electric Propulsion Technical Committee.

Acknowledgments

The ideas presented in this monograph were years in the making, and over that time the authors have benefited from numerous conversations with many colleagues. The authors were extremely fortunate that they had the opportunity to gain insights through discussions with luminaries of the field, the late Profs. Ralph Lovberg and Robert Jahn. The authors have also benefited from many extremely helpful conversations on this topic over the years with Dr. Adam Martin, Dr. Michael LaPointe, Mr. J. Boise Pearson, Dr. David Kirtley, Prof. Pavlos Mikellides, and Prof. Edgar Choueiri.

The authors would like to thank Mr. Curtis Promislow for his aid in the generation of Fig. 4.1 and for capturing the photograph on the cover of this monograph.

CHAPTER 1

Introduction

D EVICES and machines that couple with and accelerate objects through a magnetic field have become ubiquitous in our lives over the last century. Examples can be found in a wide variety of applications, from everyday electric induction motors to high-tech maglev trains, from in-space plasma thrusters to pumps for conducting fluids, from particle accelerators to fusion reaction drivers.

In this monograph, we describe a modeling technique and how it applies to specific types of accelerators that are inductively-coupled to conducting media, driving currents in the media through the application of time-varying magnetic fields. This interaction is governed by a combination of Ampère's law and Faraday's law,

$$\nabla \times \mathbf{B} = \mu \mathbf{j}, \tag{1.1a}$$

$$\nabla \times \mathbf{E} = -\frac{\partial \mathbf{B}}{\partial t}, \tag{1.1b}$$

where $\mathbf{B}$ and $\mathbf{E}$ are the magnetic and electric field vectors, respectively, $\mathbf{j}$ is the current density, and μ is the magnetic permeability. Ampère's law describes how electrical current produces a magnetic field. If the current is time-varying, it will produce a time-varying magnetic field that, in turn, produces an electric field in accordance with Faraday's law. The electric field induces a current in the body to be accelerated, which subsequently interacts with the magnetic field to yield a $\mathbf{j} \times \mathbf{B}$ Lorentz body force that performs electromagnetic work.

Inductively-coupled accelerators, while sharing many similarities with directly-coupled accelerators, such as railguns or pulsed plasma thrusters, are different in distinct ways. While the moving media complete the circuit in directly-coupled accelerators, the external drive circuit in an inductive accelerator is completed independent of the moving, accelerated body. As such, current in the external drive circuit of an inductive accelerator can flow without any moving media present. In addition, the current in the directly-coupled circuit is in line with and driven by the external power source, while the current in the moving inductively-coupled media is driven by time-varying fields in the manner described above. This makes time variation of the current in the external drive circuit necessary and essential to the inductively-coupled acceleration process. Finally, if current in the external circuit is flowing without a movable, conducting media present, then the energy driven in the external circuit will simply be resistively dissipated without performing any useful work.

1.1 TAXONOMY OF INDUCTIVE-COUPLED ACCELERATORS

The time-varying fields found in inductively-coupled accelerators can be produced by one of two general means. While there are variations and specializations on both, the typical means of production are either with a pulsed circuit or through the application of a sinusoidal alternating current (AC). A taxonomy of inductively-coupled accelerators and how they subdivide into each mode of field production is given in Fig. 1.1.

The present monograph focuses on circuit modeling of the pulsed accelerators enclosed in the dashed box. While there are numerous electrical means to produce a fast current pulse in the external driving circuit, the easiest means is to rapidly

Inductively-Coupled Accelerators

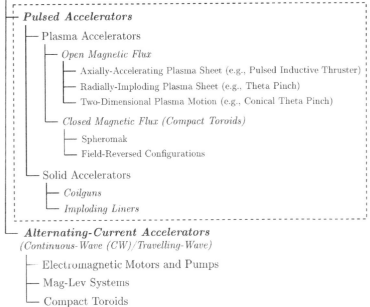

Figure 1.1 Taxonomy of inductively-coupled accelerators with the subject of this monograph in the dashed box.

discharge a capacitor through an inductive element, such as a winding or coil, producing a time-varying field that interacts with the accelerated medium. The medium being accelerated may be either a plasma or a conducting solid (though one could envision instances where a conducting liquid could replace the solid, or where the material will begin as a solid or liquid, and upon the application of a great current the material melts, vaporizes, and ionizes to yield a conducting gas).

Inductive pulsed plasma accelerators [1] are typically categorized by the nature of their magnetic flux lines. While all magnetic flux lines close on themselves, the term "open" magnetic flux in this case refers to a configuration where the currents in the external driving circuit coil and the plasma

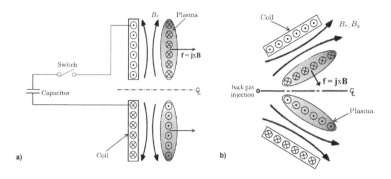

Figure 1.2 Schematics showing the basic operation of an open magnetic flux: a) axially-accelerating plasma sheet and b) two-dimensional conical theta pinch accelerator, where the Lorentz body force $\mathbf{f}$ arises from the interaction between the azimuthal plasma current density $\mathbf{j} = -j_p\hat{\theta}$ and the magnetic field, which generally has radial and axial components $\mathbf{B} = B_r\hat{r} + B_z\hat{z}$ (from [1]; licensed under a Creative Commons Attribution (CC BY) license).

produce a concentrated axisymmetric r-z magnetic field sandwiched between the two currents. This can be accomplished either in the planar configuration shown in Fig. 1.2a, producing an axially-accelerating plasma sheet, or in the conical theta-pinch configuration shown in Fig. 1.2b, yielding a plasma sheet that is accelerated both in the radially-inward and axial directions. In contrast to either of those devices, the plasma in a third design permutation known as the theta-pinch is inductively squeezed in a pure radially-inward direction. In any of these configurations the concentrated field lines in the r-z plane, once leaving the space between the external coil and the plasma, expand to fill space until they reconnect to themselves on the other side of the external coil as illustrated in Fig.1.3.

In "closed" magnetic flux accelerators, the induced current produces a closed magnetic field structure that is embedded

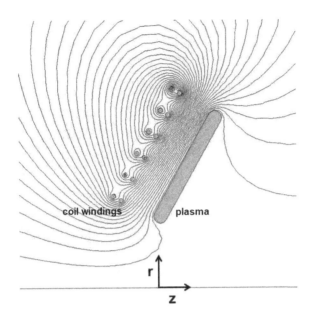

Figure 1.3 Calculated magnetic flux contours for a coil with six two-turn leads in parallel, a cone angle of 60°, and a plasma thickness of 1 cm (from [1]; licensed under a Creative Commons Attribution (CC BY) license).

in the plasma. These combined magnetic field-plasma structures, known as plasmoids or compact toroids, isolate the plasma from its surroundings and aid in plasma cohesion and shape retention while under acceleration. Compact toroids generally fall into two categories, as illustrated in Fig. 1.4: spheromaks possessing both poloidal and toroidal magnetic fields, and field-reversed configurations (FRC) having only poloidal fields.

The accelerated mass in an inductively-coupled accelerator need not be a plasma. There are many other solids (and liquids and gases) that will support currents induced by an externally-applied time-varying magnetic field. Solid conducting bodies, for example, typically possess electrical

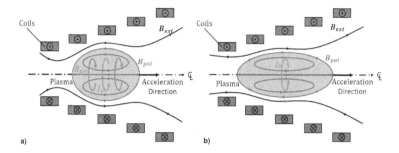

Figure 1.4 Schematic illustrations of conical coil sets containing a) a spheromak plasma and b) a field-reversed configuration (FRC) plasma, showing the directional sense of the current in the coils, the azimuthal plasma current j_p, and poloidal magnetic field B_{pol}, the magnetic field external to the plasma B_{ext}, and the toroidal magnetic field B_{tor} in the spheromak configuration (from [1]; licensed under a Creative Commons Attribution (CC BY) license).

conductivity values that are several orders of magnitude greater than a plasma; and beyond the skin depth, the induced currents are extremely effective at excluding the magnetic field from the bulk of the material. The coilgun, shown schematically in Fig. 1.5, involves an inductively-coupled projectile (sometimes called a macro-particle or macron) that is axially accelerated as the magnetic field preferentially collapses radially-inward behind it. In the case of an imploding liner, an initially-solid liner is forced to collapse radially inward. This is accomplished either through the application of electromagnetic forces or through the melting/vaporization of the material. The solid accelerators are distinct from the plasma accelerators in that the material structure of the former either maintains its shape (as in the case of a coilgun) or must be overcome to radially implode the conductor (as in the case of liners).

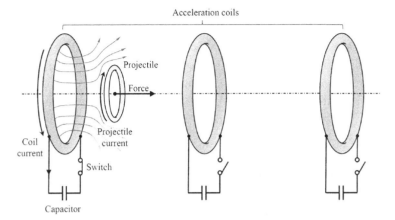

Figure 1.5 Schematic illustration of a multi-stage coilgun accelerator.

The class of accelerators that establish a continuous-wave (CW) or travelling wave magnetic field structure to impart momentum to conducting or magnetically-permeable bodies is beyond the scope of this monograph, but for the sake of completeness we take a moment here to describe their operation. CW/travelling wave systems are fundamentally different in operation relative to pulsed systems because of the typical timescales involved, with changes in the magnetic field of pulsed systems often occurring orders of magnitude more rapidly. This disparity of timescales allows a greater degree of magnetic field diffusion into the accelerated medium in CW/travelling wave systems. These systems typically possess one or more sets of coils that are driven by a sinusoidal AC current waveform (or several current waveforms offset in phase by some fraction of 2π radians). The magnetic field vector rotates about an axis or translates linearly downstream as the successive coils are sequentially energized and de-energized, inducing currents in the conducting medium. This process has been used to form and accelerate compact toroid/FRC plasma structures. The process described is also the basis for all heavy electric motors, and the inverse process is representative of how an electrical power generator operates. Induced currents in travelling wave systems can be configured to continuously accelerate or pump a conducting medium, as found in maglev trains, liquid metal pumps, and various plasma accelerators. In all instances, the currents in the medium interact with the magnetic field to produce or sustain motion.

1.2 DOMAIN OF THE PRESENT WORK

While the types of accelerators discussed and modeled in this text are pulsed, inductively coupled units, there are several other means to electromagnetically accelerate a medium through its interaction with a time-varying magnetic field that are beyond the scope of the present work. The CW/travelling wave accelerators, listed in the inductively-coupled accelerator taxonomy and briefly described above, have an operating mode and commensurate analysis that are fundamentally different from the pulsed accelerators. Also not covered in this text are pulsed directly-coupled accelerators, including devices such as railguns and pulsed plasma thrusters. Despite some similarities in the models, directly-coupled and inductively-coupled accelerators have a fundamental difference. In the former the moving body (armature) completes the primary circuit, while in the latter the external circuit is completed irrespective of the presence of the moving body, which can only couple to the external circuit through the presence of a magnetic field. Though we will not discuss magnetic nozzles explicitly, they can be analyzed within the modeling framework when deployed in conjunction with an inductively-coupled accelerator. Finally, there are devices that use magnetic fields (both steady-state or time-varying) to direct energy into a body or a flow but where the acceleration mechanism does not arise due to an electromagnetic Lorentz force. Examples of these include applications employing radio-frequency or microwave power deposition, magnetic nozzles, or acceleration through a double-layer or pondero-motive force.

1.3 APPLICABILITY OF THE MODELING TECHNIQUE

A lumped-element circuit model coupled to the momentum equation is used in this monograph to calculate the time-history of the current flowing in the various electrical circuits and the resulting motion of the accelerated body. Axisymmetry is assumed in the calculation of the one- or two-dimensional motion of the body in the r-z plane. The body itself is treated as a single element modeled in a Lagrangian reference frame. The physical and electrical connection between the moving body and any external circuits are described by the mutual inductance between the various inductive circuit elements.

It should be understood that while the modeling technique described in this monograph can be tailored to represent relatively sophisticated systems, it still assumes that the body is a single highly-conductive element moving in one- or two-dimensions in the r-z plane, and that the circuit elements can all be represented using lumped-element equivalent models. The model is not intended to capture all the phenomena that might be observed using more detailed two- or three-dimensional fluid or particle-in-cell (PIC) techniques techniques. The simplicity of the model allows relatively straightforward application to a wide variety of pulsed inductively-coupled accelerators, facilitating interpretation of trends in experimental data and the derivation of useful performance scaling relations. The present work discusses the use of the model in various situations, but it is noted that there are only a handful of instances where sufficient data exist to permit a discussion related to validation of the modeling techniques. The only type of accelerator for which a large enough data set exists to permit broad-based comparisons with modeling results is the open magnetic flux axially-accelerating plasma sheet [2]. For this application, the model has exhibited remarkable success in capturing performance scaling as a function of controllable parameters, quantitatively matching performance data, and uncovering insights that align with those observed using more sophisticated modeling methods.

There has been some effort on modeling acceleration in conical theta pinch two-dimensional plasma accelerators [3, 4]. For this application, there is minimal data for quantitative high-fidelity comparisons. The trends captured by the model appear to qualitatively match the available data and capture some of the same general trends, but there are still significant knowledge gaps that prevent more global validation of the modeling techniques for two-dimensional accelerators. A quantitative comparison to existing data is further complicated by the fact that in experiments the plasma may have only been partly ionized, violating the "highly-conductive" assumption and allowing for excessive diffusion of the magnetic field through the plasma without performing appreciable magnetic acceleration.

There has been significant discussion on use of the lumped-element modeling techniques for coilguns [5, 6], and to a lesser extent various compact toroids and rotating magnetic field configurations with and without applied static background magnetic fields [7, 8, 9]. However, for these cases experimental data that would permit modeling comparisons are limited and validation of the modeling techniques for these configurations remains an open area of research.

CHAPTER 2

Fundamental Concepts for Inductively-Coupled Pulsed Acceleration

PRIOR to discussing the modeling techniques that are the subject of this monograph, we first introduce a number of terms that will be useful in developing an understanding of inductively-coupled pulsed accelerators. The remainder of this chapter is dedicated to introducing and defining fundamental terms and concepts that will arise in this text.

2.1 RESISTANCE

Electrical resistance R is probably the most familiar term in circuit analysis. It is an intrinsic property of an object (or more generally, a conduction pathway) that relates the electrical potential or voltage V required to drive a particular value of current I through the object (or pathway). A conductor with a greater resistance will require a greater potential to drive the same level of current through it. The relationship between voltage, current, and resistance is given by the simple and familiar equation

$$V = IR. \tag{2.1}$$

2.2 CAPACITANCE AND STORED ELECTRICAL ENERGY

In the context of inductively-coupled pulsed accelerators, electrical energy is typically stored in a capacitor charged to high voltage. Equal and opposite amounts of charge are located on the two conducting plates of the capacitor, with electrical energy stored in the electric field between the plates. The capacitance C is an intrinsic property of the capacitor that quantifies the ability to store electrical energy within the electric field between the plates. It is a function of the area of the plates, their separation distance, and the electrical permittivity of the separating material. For a capacitor, the stored energy is proportional to the capacitance and the square of the voltage V_0 between the two sides, and is written as

$$E_C = \frac{1}{2}CV_0^2, \tag{2.2}$$

where E_C is the stored energy. In the course of designing an inductively-coupled pulsed accelerator, both C and V_0 can be varied. However, the values of these parameters affect the dynamics of the current pulse, which can in turn impact the performance of the accelerator.

2.3 INDUCTANCE AND MAGNETIC FIELD ENERGY

Inductance L is a property of any current-carrying material that is solely related to the geometry of the path followed by the current flowing through the material and the magnetic permeability of the space surrounding the material. As discussed in the previous chapter, flowing currents will produce magnetic fields in the manner prescribed by Ampère's law in Eq. (1.1a). The amount of energy stored in the magnetic field E_M owing to a current flowing through an isolated conductor is directly related to the self-inductance of the conductor L and the square of the flowing current. This "self" magnetic field energy is written as

$$E_{M\,\text{self}} = \frac{1}{2}LI^2. \tag{2.3}$$

If there are multiple independent conductors, each will have a self-inductance and associated "self" magnetic field energy. When multiple current loops have current flowing in them at the same time, the magnetic vector fields for each conductor combine according to the principle of linear superposition, giving rise to an additional magnetic field energy component associated with the pairwise mutual interaction between the conductors. This "mutual" magnetic field energy is written as

$$E_{M\,\text{mutual}} = M_{12}I_1I_2, \tag{2.4}$$

where M_{12} is the mutual inductance between conductors 1 and 2 ($M_{12} = M_{21}$) and I_1 and I_2 are the respective currents in each conductor.

Another way to think about inductance is in relation to the level of magnetic flux produced when current flows through the conductor. The magnetic flux is the integral of the magnetic field perpendicular to the surface bounded by the conducting path, and it can be written as

$$\Phi = \int_S \mathbf{B} \cdot d\mathbf{S}. \tag{2.5}$$

The inductance can then be defined as in terms of the magnetic flux as

$$L = \frac{\Phi}{I}. \tag{2.6}$$

Changes in current will give rise to an electric field through Faraday's law given in Eq. (1.1b), which will produce an electromotive force in the conductor manifesting as an additional voltage that opposes the change in current. This inductive voltage V_L can be written in terms of the self-inductance and the rate of change of the current as

$$V_L(t) = L\frac{dI}{dt}. \tag{2.7}$$

The mutual inductance quantifies the reaction of an independent current path (loop 1) to a time varying magnetic flux appearing inside loop 1 but created by a separate current path (loop 2). This gives rise to an inductive voltage V_M in loop 1 that is a function of the mutual inductance and the rate of change of the current in loop 2, and can be written as

$$(V_M)_1(t) = M_{12}\frac{dI_2}{dt}. \tag{2.8}$$

This response is the basis for transformers and other systems where the responses of two conductors are coupled through a time-varying magnetic field.

2.4 VIRTUAL WORK AND ELECTROMAGNETIC FORCES

Using the definitions of the magnetic field energy written in the previous section, we can employ the principle of virtual work to connect the changing magnetic field energy as a function of conductor motion to the axial and radial Lorentz body forces on the conductor. Assume there are two conductors (1 and 2) and conductor 2 can move in space, capable of translation in z and variation of its radius r. The magnetic field energy of the system is

$$\begin{aligned}
U_B &= (E_{M\text{self}})_1 + (E_{M\text{self}})_2 + (E_{M\text{mutual}})_{12}, \\
&= \frac{1}{2}L_1I_1^2 + \frac{1}{2}L_2I_2^2 + M_{12}I_1I_2. \tag{2.9}
\end{aligned}$$

For a differential displacement of conductor 2 in the r or z direction, the corresponding differential change in the magnetic field energy is

$$\delta U_B = \left(\frac{1}{2} \frac{\partial L_2}{\partial r} I_2^2 + \frac{\partial M_{12}}{\partial r} I_1 I_2 \right) \delta r + \frac{\partial M_{12}}{\partial z} I_1 I_2 \delta z. \quad (2.10)$$

By definition, a force applied in the direction of an infinitesimal displacement δx will perform work to change the energy of the system as $\delta U = F_x \, \delta x$. Applying this relation to Eq. (2.10) results in axial and radial Lorentz forces given as

$$\begin{aligned} F_z &= \frac{\partial M_{12}}{\partial z} I_1 I_2, \\ F_r &= \frac{1}{2} \frac{\partial L_2}{\partial r} I_2^2 + \frac{\partial M_{12}}{\partial r} I_1 I_2. \end{aligned} \quad (2.11)$$

2.5 PERFORMANCE METRIC DEFINITIONS

In any electrical accelerator, performance can be quantified through a measure of the maximum velocity attained by the body being accelerated and the efficiency of the process by which the body arrived at that velocity. For any pulsed device, we can define a specific impulse (thrust per unit weight flow of ejected material) based upon the maximum achieved exit velocity u_e. Normally the exhaust velocity will be the z-component of the velocity u_z, as this value in an axisymmetric system is tied to the amount of momentum transferred during a pulse. This can also be set equal to a ratio containing the impulse bit I_{bit} or impulse per pulse and the mass bit m_{bit} or mass expended or ejected per pulse. Using customary definitions, the specific impulse I_{sp} can be written as

$$I_{\text{sp}} = \frac{u_e}{g_0} = \frac{I_{\text{bit}}}{m_{\text{bit}} g_0}, \quad (2.12)$$

where g_0 is the gravitational acceleration constant for Earth.

The efficiency, defined as the ratio of the ultimate kinetic energy attained by the accelerated body divided by the initially stored electrical energy E_0 or discharge energy per pulse is written as

$$\eta = \frac{m_{\text{bit}} u_e^2}{2E_0} = \frac{I_{\text{bit}}^2}{2m_{\text{bit}} E_0}. \qquad (2.13)$$

CHAPTER 3

Electrical Circuits

W HEN applying circuit modeling techniques to inductively-coupled pulsed accelerators it is important to create an accurate representation of the accelerator as a network of individual circuit elements, and to capture not only the behavior of each individual circuit element but also the manner in which those elements interact with each other through the electromagnetic field. The interaction can appear complex to model, since multiple individual circuits can simultaneously interact with each other. This is in addition to the fact that the motion of the accelerated body through space always gives rise to a time-varying inductive element in the formulation.

To introduce the manner in which such circuits can be modeled, we initially consider the very simple resistive-inductive-capacitive (RLC) circuit shown in Fig. 3.1. Applying Kirchoff's voltage law to the circuit and including a statement for the conservation of charge on the capacitor, we can write

$$\frac{d}{dt}\left(IL\right) + RI - V_{\mathrm{C}} = 0 \qquad (3.1a)$$

$$\frac{dV_{\mathrm{C}}}{dt} = -\frac{I}{C} \qquad (3.1b)$$

where I is the current in the circuit, L is the inductance, R is the resistance, and V_{C} is the time-varying voltage on a

DOI: 10.1201/9780429351976-3

capacitor with capacitance C. If L is constant, these equations can be combined and reduced to a single second-order ordinary differential equation for the current flowing through the system. The solution to this equation mirrors the damped sinusoidal response of an oscillating spring-mass-damper system, where energy is resistively dissipated as it oscillates between the capacitive (electric field potential energy storage) and inductive (magnetic field potential energy storage) elements. Most importantly, no useful work is performed in this type of circuit where L is constant.

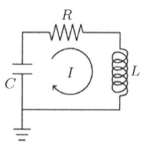

Figure 3.1 Circuit topology for a simple RLC circuit.

For the case where L is time-varying, Eq. (3.1a) can be rearranged and multiplied by the current, resulting in an equation stating how power delivered by the capacitor is subdivided between the resistive and inductive elements of the circuit:

$$P = V_{\mathrm{C}}I = I^2 R + LI\frac{dI}{dt} + I^2\frac{dL}{dt}. \tag{3.2}$$

This can also be recast as

$$P = V_{\mathrm{C}}I = I^2 R + \frac{d}{dt}\left(\frac{LI^2}{2}\right) + \frac{I^2}{2}\frac{dL}{dt}, \tag{3.3}$$

where the first two terms on the right-hand side are commonly recognized as the resistive power dissipation and the time rate of change of the energy stored in the inductor, respectively. The third term represents the electromagnetic work performed on the moving conductor, showing how the time-varying inductance associated with the moving, accelerated body is critical in converting stored electrical energy into useful work.

3.1 GENERAL ELECTRICAL CIRCUITS FOR INDUCTIVELY-COUPLED PULSED ACCELERATORS

Having introduced the idea that a time-varying inductive element is required to perform work in an RLC circuit, we proceed with the modeling of a general pulsed accelerator consisting of multiple independent external circuits that all couple inductively to one another and to a moving conductor. The interacting circuits can be considered as axisymmetric coils of wire while the moving body is an axisymmetric conducting medium of any type that supports the flow of an azimuthal current. We assume that at a given time, current flows in coils l through m. The coils outside of this range are assumed to be open circuits with infinite impedance that, consequently, do not interact with any conductors considered in the problem. The terms l and m are used, as opposed to 1 through m, because at some point during the acceleration event rapid turn-off switches or blocking diodes may be used to interrupt the current flow in some of the coils. In addition, switches may be activated some time after $t = 0$, allowing current to flow in additional coils as their effects are added to the overall system. We recognize that as the coils are switched on and off, the active coils may not necessarily be sequential in number (for example, the active coils may be numbers 3, 4, 7, and 9). The notation adopted is only to make the mathematical representation easier and more compact to write and is in no way intended to exclude non-sequential coil combinations from simultaneous operation.

Figure 3.2a shows a general lumped-element circuit representation of the i-th external coil circuit (for $l \leq i \leq m$), while the corresponding circuit representation of the moving conductive body p is shown in Fig. 3.2b. The current arrows indicate the direction of positive current flow. This is done so azimuthally flowing currents in different conductors will always have the same sign when moving in the same direction in physical space.

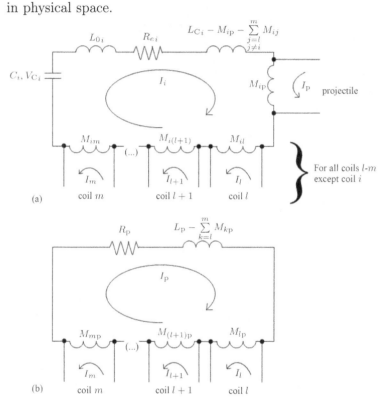

Figure 3.2 a) General circuit schematic for interactions between the i-th external circuit, the moving body p, and all other active external circuits. b) Circuit schematic showing the interactions between the moving body p and all active, external circuits.

The i-th external circuit is powered by a discharging capacitor having a time-varying voltage V_{Ci}. The circuit is modeled as having a lumped element resistance R_{ei} and a self-inductance comprised of the circuit stray inductance L_{0i}

and the inductance of the coil $L_{\mathrm{C}i}$. The circuit interacts with the moving conductive body through the mutual inductance $M_{i\mathrm{p}}$, which is a function of the relative positions of the two conductors and varies in time due to the motion of the body. It also interacts with the j-th external circuit through a mutual inductance M_{ij}, which is fixed for the given geometry. It is relatively straightforward to apply Kirchoff's voltage law to the i-th circuit to compactly write its circuit equation and a statement for the conservation of charge on the capacitor as

$$V_{\mathrm{C}i} = R_{\mathrm{e}i}I_i + (L_{0i} + L_{\mathrm{C}i})\frac{dI_i}{dt} + \frac{d}{dt}(M_{i\mathrm{p}}I_\mathrm{p}) + \sum_{\substack{j=l \\ j \neq i}}^{m} M_{ij}\frac{dI_j}{dt}, \quad (3.4\mathrm{a})$$

$$\frac{dV_{\mathrm{C}i}}{dt} = -\frac{I_i}{C_i}. \quad (3.4\mathrm{b})$$

The first term on the right-hand side of Eq. (3.4a) is the resistive dissipation in the circuit, while the second term combines the effects of the stray and self-inductance of the i-th circuit. The third term represents the interaction between the i-th circuit and the moving body p while the final term is a summation of the interactions due to mutual inductance between the i-th circuit and all the other external circuits.

In addition to equations for all the external circuits, we also require one to govern the flow of current in the moving conductive body. Applying Kirchoff's voltage law, we can write this equation as

$$R_\mathrm{p}I_\mathrm{p} + \frac{d}{dt}(L_\mathrm{p}I_\mathrm{p}) + \sum_{k=l}^{m} \frac{d}{dt}(M_{k\mathrm{p}}I_k) = 0, \quad (3.5)$$

where the first term represents resistive dissipation, the second term represents the effects of self-inductance, and the final term is the summation of the interactions due to mutual inductance between the moving conductive body and all the active external circuits. As with Eq. (3.3), it is essential to understand how the time-varying self and mutual inductance terms evolve since they are intimately connected to the level of electromagnetic work performed on the moving body.

In this monograph, we assume that the conductors can be represented by their lumped-element circuit values. The self and mutual inductance terms in Eqs. (3.4a) and (3.5) are a function of the geometry of the problem. More specifically, these terms are a function of the distribution of current in each conductor. Under the assumption of axisymmetry, the self-inductance of the moving conductor can change as a function of time if the current path is compressed or expanded in the radial direction. The mutual inductance between each pairwise-interacting conductor is a function of the distribution of currents in each conductor and their relative positions. For the interactions between the moving conductor and the external circuits, it is important to recognize that the mutual inductance between these circuits will vary as the conductor moves in the axial and radial directions. The time derivatives of the moving conductor self-inductance $L_{\mathrm{p}}(r)$ and the mutual inductance $M_{k\mathrm{p}}(r, z)$ can be written as:

$$\frac{dL_{\mathrm{p}}}{dt} = \frac{\partial L_{\mathrm{p}}}{\partial r} v_r, \qquad (3.6a)$$

$$\frac{dM_{k\mathrm{p}}}{dt} = \frac{\partial M_{k\mathrm{p}}}{\partial r} v_r + \frac{\partial M_{k\mathrm{p}}}{\partial z} v_z, \qquad (3.6b)$$

where v_r is the radial rate of compression or expansion of the moving body while v_z is the axial velocity.[1]

While at least one external circuit must be time-varying to induce a current in the moving body, it is interesting to note that there will be electromagnetic interactions between the body and the other external circuits even if the currents in those external circuits are not time-varying. This interaction

---

[1] The inductance of the current path also changes as the current density configuration changes (for example, as the current path widens in the axial direction through resistive diffusion), but this variation is often assumed to be small over the duration of an accelerating pulse. In this monograph, the inductance as a function of current position and configuration is typically calculated *a priori* using a separate magnetic field solver. The introduction of a variable current density distribution, while possible to include, adds to the computational difficulty of representing the accelerating body as a lumped element.

arises from the time-varying mutual inductance term, which will change according to Eq. (3.6b) due to the relative motion between the moving body and each external circuit.

This text will primarily use Eqs. (3.4), which are written for simple pulsed RLC circuits powered by discharging capacitors. However, it is possible to re-write and tailor this set of equations to represent nearly any arbitrary current source (DC, continuous-wave, programmable ramp, square-wave, pulse-forming network, etc.) so long as the response of that source to the time-varying impedance can be properly represented. Illustrations of some of these more complex sources can be found in Refs. [1, 2] and the references therein. The example of a constant DC current could be particularly useful in instances where a coil of copper wire or superconducting material is used to generate an applied static magnetic field, such as in the case of an applied DC magnetic bias field or magnetic nozzle. Another permutation not covered in this text is the modeling of a system where there are multiple moving conductive bodies. This would necessitate the addition of terms to Eq. (3.4a) to represent the interaction between the i-th coil and these additional bodies, the inclusion of additional instances of Eq. (3.5) to model the current flowing in those additional moving conductors, and the addition of terms in Eq. (3.5) to model the pairwise interactions between moving conductors.

3.2 CALCULATING SELF AND MUTUAL INDUCTANCE

Linear superposition of magnetic fields permits us to write the overall magnetic field configuration for external circuits $l \leq j \leq m$ and moving conductor p as

$$\mathbf{B}(r, z) = \sum_{j=l}^{m} \mathbf{B}_j(r, z) + \mathbf{B}_{\mathrm{p}}(r, z), \qquad (3.7)$$

where $\mathbf{B}_j$ is the magnetic field arising from current flowing in the j-th external circuit and $\mathbf{B}_{\mathrm{p}}$ is the magnetic field produced by the current in the moving body. As an illustrative example, we can write the magnitude of the magnetic field for a system possessing three external circuits ($l = 1$, $m = 3$) as

$$
\begin{aligned}
|\mathbf{B}|^2 &= |\mathbf{B}_1|^2 + |\mathbf{B}_2|^2 + |\mathbf{B}_3|^2 + |\mathbf{B}_\mathrm{p}|^2 \\
&+ 2\mathbf{B}_1 \cdot \mathbf{B}_2 + 2\mathbf{B}_1 \cdot \mathbf{B}_3 + 2\mathbf{B}_2 \cdot \mathbf{B}_3 \\
&+ 2\mathbf{B}_1 \cdot \mathbf{B}_\mathrm{p} + 2\mathbf{B}_2 \cdot \mathbf{B}_\mathrm{p} + 2\mathbf{B}_3 \cdot \mathbf{B}_\mathrm{p}.
\end{aligned} \tag{3.8}
$$

The commensurate magnetic field energy associated with this configuration is

$$
\begin{aligned}
E_\mathrm{M} &= \frac{L_1 I_1{}^2}{2} + \frac{L_2 I_2{}^2}{2} + \frac{L_3 I_3{}^2}{2} + \frac{L_\mathrm{p} I_\mathrm{p}{}^2}{2} \\
&+ M_{12} I_1 I_2 + M_{13} I_1 I_3 + M_{23} I_2 I_3 \\
&+ M_{1\mathrm{p}} I_1 I_\mathrm{p} + M_{2\mathrm{p}} I_2 I_\mathrm{p} + M_{3\mathrm{p}} I_3 I_\mathrm{p}.
\end{aligned} \tag{3.9}
$$

The first row of terms on the right-hand side of the equation represents the magnetic field energy contributions owing to the self-inductance of each independent conductor, while the second and third rows are the contributions owing to the mutual inductance that arises through pairwise interactions between independent conductors (note: $M_{xy} = M_{yx}$).

The self-inductance terms can be measured or they can be found using a finite element model to calculate the resulting magnetic field that arises from current flowing in each conductor while in isolation. Using finite element analysis, the magnetic field energy $(E_\mathrm{M})_j$ of the j-th conductor in isolation is found by multiplying the magnetic energy density $\varepsilon_B = |\mathbf{B}|^2 / (2\mu)$ of each cell by the cell volume and then summing all contributions over the entire domain. The self-inductance of the j-th conductor is then

$$
L_j = \frac{2 \, (E_\mathrm{M})_j}{I_j{}^2}. \tag{3.10}
$$

The value of the magnetic field linearly scales with the current through Ampère's law, so dividing by the square of current in Eq. (3.10) removes from the self-inductance any dependence on the current.

Once the self-inductance of each conducting element is known, the mutual inductance terms can be determined for each pairwise combination of conductors. Again using finite element analysis, current is applied to the set of pairwise conductors of interest and the magnetic field energy is calculated as before by summing the magnetic energy density contributions over the computational domain. For the example of elements 1 and 2, this leads to a total magnetic field energy of

$$E_{\mathrm{M}} = \frac{L_1 I_1{}^2}{2} + \frac{L_2 I_2{}^2}{2} + M_{12} I_1 I_2. \qquad (3.11)$$

The self-inductance values L_1 and L_2 were previously found using Eq. (3.10) and the current values I_1 and I_2 are set by the user, so this equation can be easily solved to find the value of M_{12}. As noted in the previous section, the mutual inductance is a function of the relative current distributions in pairwise-interacting conductors, so the calculations must be revisited if the geometry of the problem is altered. Also, while the mutual inductance is fixed for interactions between external circuits, it varies for interactions between the individual external circuits and the moving conductor. As such, for a moving conductor the calculation must be performed many times to assemble a tabulated set of mutual inductance values as a function of conductor position. It is straightforward to use a table lookup method, interpolating values of $M_{kp}(r, z)$ and calculating the gradients in mutual inductance using a finite difference method.

It should be noted that there are several assumptions buried in the previous method of calculating inductance, and it is not easy to calculate the "true" inductance for each element in the system using simple magnetic field finite element solvers. The astute observer will note that the distribution of current in one conductor will be affected by the current flowing in every other conductor in the system. It can also be affected by the skin effect in the various conductors, with the current penetration depth in each conductor being a function of the frequency of the current and the conductivity. Even if the finite element solver being used allows the user to specify the current distributions in every conductor (as opposed to the solver determining this for each isolated inductor and pairwise-interacting set), the distribution is also affected by the position of the moving conductor in the system and the relative magnitudes and frequencies of the currents in each conductor, which are generally not known *a priori*. These factors will slightly alter the self and mutual inductance values. While the form and magnitude of the calculated solution are not greatly affected, we shall show later in this monograph that even small deviations in the inductance values owing to variations in the current density distribution can lead to differences in the computed current waveforms and the ultimate velocity. The latter can vary in excess of 10% depending on the assumed current distribution. Further refinement of the interactions between conductors and elimination of some of the errors introduced by the assumptions in this section dip into the realm of coupled 2- and 3-D analysis of the problem, moving away from the simplified but insightful lumped element modeling framework that is the subject of this monograph.

In instances where there is only one external coil and one moving conductor, it is possible to make some additional simplifying assumptions in writing the self and mutual inductance terms. This adds an additional yet slight deviation from the "true" values, but it further simplifies the problem. We proceed with a short discussion of how this has been done in the past to simplify the problem.

3.2.1 One Coil, One Body Mutual Inductance

For a single external coil (1) coupled to a moving conductive body (p) with a circuit shown in Fig. 3.3a, the circuit equations from Eqs. (3.4) and (3.5) are written as

$$V_{C1} = R_{e1}I_1 + (L_{01} + L_{C1})\frac{dI_1}{dt} + \frac{d}{dt}(M_{1p}I_p), \quad (3.12a)$$

$$R_pI_p + \frac{d}{dt}(L_pI_p) + \frac{d}{dt}(M_{1p}I_1) = 0, \quad (3.12b)$$

$$\frac{dV_{C1}}{dt} = -\frac{I_1}{C_1}. \quad (3.12c)$$

As before, the directions assumed for the currents are such that in physical space currents moving in the same direction will have the same sign.

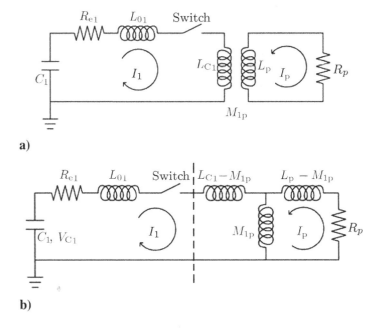

a)

b)

Figure 3.3 a) General lumped-element circuit representation and b) equivalent circuit for an inductively-coupled pulsed accelerator consisting of a single external coil (1) coupled to a moving conductive body (p).

We can redraw the circuit in the manner shown in Fig. 3.3b. If we assume the inductive reactance is much greater than the plasma resistance, then we can combine the inductive elements to the right of the dashed line to yield a Thévenin-equivalent inductance L_{th}. This results in the following equation for the total inductance of the circuit (including the stray inductance L_{01})

$$L_{tot} = L_{01} + L_{th} = L_{01} + \left(L_{C1} - \frac{M_{1p}^2}{L_p} \right). \qquad (3.13)$$

It is straightforward to isolate and measure the coil self-inductance L_{C1} by connecting a meter that measures inductance at the location of the dashed line in Fig. 3.3b. It is not as simple to quantify the self-inductance of the moving body L_p and if that body is collapsing inwardly, as one might find in the case illustrated in Fig. 1.2b, then the self-inductance must be quantified as a function of the change in the radius of the body (and that doesn't account for the fact that the angle between the conductor and the centerline may also change as the body collapses). However, if that is known or can be determined, either through measurements or using finite element magnetic field modeling, then the mutual inductance M_{1p} as a function of body position can be found either through measurement or modeling. Assuming that the size and shape of the distributed current in the body is reasonably well known, the mutual inductance can be measured by placing a simulated body (a metallic sheet in the expected shape of current distribution of the moving body) in proximity to the external coil and then moving it in the axial direction to quantify the functional relationship with z. Additional simulated bodies of differing radii can be used to quantify the functional relationship with r. Doing this while using a meter again connected at the dashed line, we can employ Eq. (3.13) to build a map of $M_{1p}(r, z)$ at discrete points in the r-z plane.

Work by Martin *et al.* [10] showed good agreement between measurements of L_{th} and magnetic field modeling of a flat, planar coil consisting of multiple axisymmetric rings containing a time-varying current acting upon a annular disk having a finite conductivity. Hallock [3], working with a conical theta pinch geometry, showed that measurements of L_{th} aligned well with magnetostatic field simulations of the problem where the coil was again modeled as multiple axisymmetric rings and the conductor was represented as a finite-thickness edge of a frustum mirroring the coil angle and axial length and having a zero magnetic flux boundary condition. Example comparisons between the experimentally measured mutual inductance values and those found using modeling are shown in Fig. 3.4 for a flat, planar coil and a conical theta-pinch coil. This method should capture with high precision the inductance of a solid accelerated body that does not deform during the discharge. For a plasma, the method assumes a particular shape and thickness of the current channel as a function of position. Increasing deviations from the assumed current channel configuration will lead to similarly increasing errors in the computed quantities. While this does not invalidate the approach of this monograph, it does highlight one of the limitations of the lumped-element approach, especially when performing quantitative comparisons between experimental data and the results of the model.

3.2.1.1 Additional Simplifications

In work on axially-accelerating plasma sheets, Lovberg and Dailey [11, 12] assumed that L_p was not a function of time (no radial motion) and that it was equal to L_{C1}. Several works have been performed under this assumption, and it is useful to determine how close the past models are to those that use the more general formulation. Assuming $L_{C1} = L_p$, Eq. (3.13) becomes

$$L_{tot} = L_{01} + \left(L_{C1} - \frac{\overline{M_{1p}}^2}{L_{C1}} \right), \qquad (3.14)$$

where we have relabeled M_{1p} as $\overline{M_{1p}}$ to keep track of the fact that the assumptions made could result in the values of the the mutual inductance being different from those found using

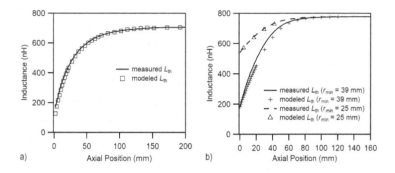

Figure 3.4   Measured and modeled profiles of L_{th} as a function of axial separation for a) a flat, planar coil and a conducting body in the shape of an annular disk (after Ref. [10]); and b) a conical theta pinch coil and a conducting surface in the shape of a frustum having a half cone angle of 20°, a height in the axial direction of 50 mm, and minor radii r_{min} as indicated in the figure (after Ref. [3]). The conical coil has the same dimensions as the frustum with a fixed minor radius of 39 mm.

the more general formulation. As in the previous discussion, the terms in the parentheses again represent the Thévenin-equivalent inductance L_{th} of the circuit. In this formulation, it was further found that for the special case of an axially-accelerating conductive sheet in an open magnetic flux device, measurements of the inductance represented by Eq. (3.14) could be fitted using the function

$$L_{\text{tot}}(z) = L_{01} + L_{\text{th}}|_{z=0} + \overline{L_{\text{C1}}}\left[1 - \exp\left(-z/z_0\right)\right], \quad (3.15)$$

where z_0 is the decoupling distance, which is a characteristic length over which the moving conductor electromagnetically decouples from the external coil. We denote the self-inductance of the coil as $\overline{L_{\text{C1}}}$ because, unlike in the previous formulations where the coil self-inductance was determined through measurement or simulation of a coil in isolation, the value here is determined using a curve fit of the measured or simulated inductance profile. For the curve fit to work properly, a constant $L_{th}|_{z=0}$ was also included as a fit parameter. This is different from the stray inductance in the circuit L_{01} and represents the inductance associated with the fact that the coil and moving conductor are never perfectly coincident. Setting Eq. (3.14) equal to Eq. (3.15) and solving for the mutual inductance, we can write the function

$$\overline{M_{1p}} = \overline{L_{C1}} \exp\left(-z/\left(2z_0\right)\right). \tag{3.16}$$

Using this simplified method necessitates modifications of Eq. (3.12) where L_{01} becomes $L_{01} + L_{th}|_{z=0}$, L_{C1} goes to $\overline{L_{C1}}$, and M_{1p} is now $\overline{M_{1p}}$. Finally, using these substitutions in Eq. (3.6b) allows us to obtain the function

$$\frac{d\left(\overline{M_{1p}}\right)}{dt} = -\frac{\overline{L_{C1}}}{2z_0} \exp\left(-z/\left(2z_0\right)\right) v_z, \tag{3.17}$$

governing the time evolution of the mutual inductance.

3.2.1.2 Example: Comparison of Methods to Find Mutual Inductance

As a means of comparing the mutual inductance values obtained using the methods of Eqs. (3.11), (3.13), and (3.16), we present results from a set of example magnetic field simulations. In this example, shown schematically in Fig. 3.5, the coil consists of nine 2 mm diameter axisymmetric windings connected in parallel. The center of each winding is given in Table 3.1. A single-turn conducting sheet annulus was also modeled, having an inner radius of 200 mm, and outer radius of 500 mm, and a thickness of 2 mm. The current distribution in the annular disk was assumed to vary with $1/r$. The axial position of the disk is taken as the z value of the face closest to the coil. A magnetostatic solver was used for all simulations, with the data produced for Eqs. (3.13) and (3.16) calculated assuming the conducting disk boundary was magnetically impermeable.

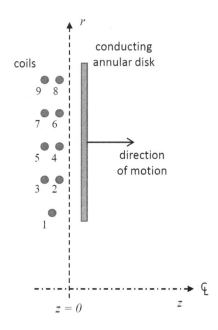

Figure 3.5 Schematic representation (not to scale) of the relative positions of the individual coil windings, for an example, single planar external coil coupled to a moving conductive body in the shape of an annular disk.

In isolation, the self-inductance of the coil was 908 nH while that of the disk was 689 nH. The mutual inductance values as a function of axial position calculated using Eqs. (3.11), (3.13), and (3.16) are shown in Fig. 3.6. In addition, the Thévenin-equivalent inductance $L_{\text{th}}(z)$ of Eq. (3.13) is also plotted in the figure. As previously noted, the value of $L_{\text{th}}|_{z=0}$ is not zero because the conducting paths have differing self-inductance values and there is a finite separation distance between the paths when the disk is at $z = 0$. We observe that the mutual inductance values calculated using Eqs. (3.11) and (3.13) are nearly coincident, while there is a slight growing offset in the value calculated using Eq. (3.16) at greater axial positions. We shall revisit this example later

Table 3.1 Coordinates for the coil windings in the planar external coil example of Fig. 3.5.

Coil Number	z [mm]	r [mm]
1	−6	202.5
2	−2.5	269.5
3	−9.5	269.5
4	−2.5	336.5
5	−9.5	336.5
6	−2.5	403.5
7	−9.5	403.5
8	−2.5	470.5
9	−9.5	470.5

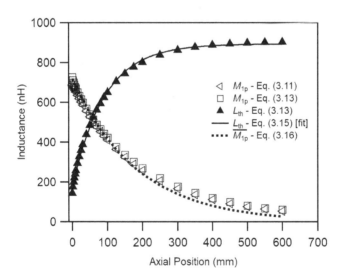

Figure 3.6 Inductance profiles calculated using the equation specified based upon magnetic field simulations of a single planar external coil coupled to a moving conductive body in the shape of an annular disk.

in this monograph to compare the calculated acceleration profiles for each case.

3.2.2 Inductance of Radially-Compressed Bodies

The techniques to quantify the self and mutual inductance of accelerated conductors as described in this section are predicated on an assumption that we know the current distributions within those conductors, their trajectories in space, and how their shape may change with time. This is straightforward when the moving conductor is a solid body that maintains its shape and can only move in one dimension, or for a conducting disk that is limited to acceleration in the axial direction. However, in the case of a compact toroid or a collapsing plasma sheet, as one might find in a conical theta pinch, the situation is more complex. While there has been some work performed on quantifying the self and mutual inductance of radially-compressed plasmas [3, 13] and compact toroids [7, 8], this method in general requires additional *a priori* assumptions regarding the path the moving conductor takes through space and the internal distribution of current as it follows that path. It is certainly possible to do this, and the results will likely be fruitful in that they will provide insight into the acceleration process, but it should be noted that for these more complex problems one should expect a reduction in the absolute accuracy of the results owing to the inherent shortcoming associated with reducing these complex conducting bodies to lumped-element circuit values.

3.2.3 Numerical Efforts of Note

Several approaches have been employed to estimate $M(r, z)$ on a discrete grid, and some are noteworthy for their distinctiveness. Novac *et al.* [14] used a 2-D numerical model for the magnetic field and used the fraction of the current that flows on the surface to calculate the mutual inductance $M(z)$ between a coil and a solid metallic projectile. Shoubao *et al.* [15] improved the current filament method using a 3-D grid-based calculation to predict the performance of a coilgun. Using a 2-D finite element model with AC magnetics and assuming a uniform current density in a plasma, Martin [4] calculated the effective inductance of the system and the self-inductance of an accelerated load and subsequently used those values to calculate $M(r, z)$ on a discrete r-z grid for conical plasma accelerators possessing specific values of the conical coil angle, θ. Using a filamentary model of current in the coils and the accelerated load, and using Green's functions to estimate the interaction between filaments, Shimazu and Slough [16] numerically calculated $M(r, z)$ of a FRC plasmoid. Alternatively, Woods *et al.* [8] sought to bridge empirical and numerical approaches by developing coupled time-dependent ordinary differential equations to calibrate a mutual inductance model in terms of non-dimensional scaling parameters.

CHAPTER 4

Equations of Motion

T HE equations of motion in the model are a description of mass, momentum, and energy conservation in the accelerated conducting body, tailored to be compatible with and coupled to the electrical circuit equations found in Chapter 3. These equations can be written to model either one- or two-dimensional acceleration in the r-z plane. We proceed with a description of the tailoring process that permits us to reduce the mass, momentum, and energy conservation equations to the point where they describe the time-evolution of these quantities for an accelerated body represented by an equivalent lumped-element. The resulting equations include representation of the inductive electromagnetic acceleration forces on the accelerated body that arise through pairwise interactions with an indefinite number of external circuits carrying either time-varying or steady-state currents.

4.1 CONSERVATION EQUATIONS

In this section, we write the general conservation equations governing the evolution of the mass, momentum, and energy in the accelerated body in differential form. We then integrate over the volume of the body, collapsing it to a single finite element moving in the r-z plane with volume-averaged, lumped-element properties.

4.1.1 Mass Conservation

For solid moving conductors (such as macron projectiles or collapsing liners), mass is conserved by simply holding it constant. The problem is more complex for the situations illustrated in Fig. 4.1 where a moving plasma encounters and potentially entrains gas (via ionizing or charge exchange collisions), which in turn adds to the overall mass of the plasma over time.

In differential form, the evolution of the mass density of the plasma can be written using the (source-free) continuity equation:

$$\frac{\partial \rho}{\partial t} + \nabla \cdot (\rho \mathbf{v}) = 0, \qquad (4.1)$$

where the mass density ρ is in general a function of r and z and $\mathbf{v}$ is the velocity vector. The first term on the left-hand side accounts for a change in the mass density in the control volume under consideration while the second term on the left represents a change in the mass density owing to a flux of mass into a control volume.

For the spatially-dependent density of any generic property y, we can integrate over the volume of the plasma $\mathcal{V}$ to transform the partial time derivative term yielding

$$\int_V \frac{\partial y}{\partial t} d\mathcal{V} = \frac{dY}{dt},$$

where Y is the lumped-element sum total of that property contained within the entire plasma. For example, if y is the mass density ρ, Y is the total mass in volume $\mathcal{V}$. Applying this to a volume integration of Eq. (4.1) results in

$$
\begin{aligned}
\frac{dm_\mathrm{p}}{dt} &= -\int_\mathcal{V} \nabla \cdot (\rho \mathbf{v}) d\mathcal{V} \\
&= -\int_\mathcal{V} (\mathbf{v} \cdot \nabla) \rho \, d\mathcal{V} - \int_\mathcal{V} \rho (\nabla \cdot \mathbf{v}) \, d\mathcal{V} \quad (4.2)
\end{aligned}
$$

where m_p is the mass of the moving body (plasma in this case). For mass accumulation, where the moving plasma

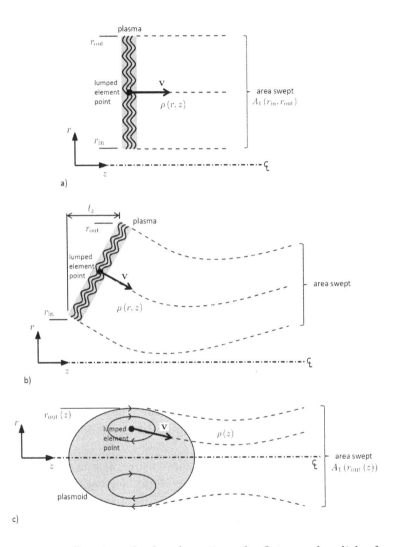

Figure 4.1 Swept paths for a) motion of a flat annular disk of plasma moving in the axial direction, b) motion of a conical plasma sheet moving in the r-z plane, and c) quasi-one dimensional axial motion of a plasmoid as it initially compresses and then expands in radius while translating axially.

volume is plowing into and entraining encountered mass, it is useful to move into a Lagrangian reference frame. This eliminates the second integral on the right-hand side and changes the sign of the first integral to be positive, because the normal to the surface area of the plasma is in the direction of motion. The integral is performed in cylindrical coordinates over the extent of the plasma to yield

$$\frac{dm_{\mathrm{p}}}{dt} = \int \int \int \left((\mathbf{v} \cdot \nabla) \rho \right) r dr \, d\theta \, dz. \qquad (4.3)$$

This integral can be rewritten as

$$\frac{dm_{\mathrm{p}}}{dt} =$$
$$\alpha \left(\int \int \int v_r \frac{\partial \rho_A}{\partial r} r dr \, d\theta \, dz + \int \int \int v_z \frac{\partial \rho_A}{\partial z} r dr \, d\theta \, dz \right),$$
$$(4.4)$$

where the terms on the right-hand side represent mass flux into the the plasma element through its radial and axial motion, respectively, into the background (ambient) neutral gas. Note that we have introduced subscript A to highlight the fact that it is the ambient neutral gas distribution that exists outside of the plasma. The coefficient α is between 0 and 1 and represents the fraction of ambient gas that is entrained by the plasma.

For the term in Eq. (4.4) representing mass accumulation through axial motion (as in Fig. 4.1a), we can integrate to yield

$$\int \int \int v_z \frac{\partial \rho_A}{\partial z} r dr \, d\theta \, dz = v_z \, A_1 \left(r_{\mathrm{in}}, r_{\mathrm{out}} \right) \rho_A \left(r, z \right)$$
$$= v_z \, \rho_{A,z} \left(r, z \right), \qquad (4.5)$$

where A_1 is the cross-sectional area of a disk in the r-θ plane having inner and outer radii r_{in} and r_{out}, respectively, and advancing into the gas in the z-direction at the rate v_z. If the plasma is moving in both the radial and axial directions, then values of r_{in} and r_{out} can themselves be functions of the time-changing plasma location (r, z). Similarly, the density is evaluated at the location (r, z) of the plasma element for that moment in time. Combining A_1 and $\rho_A(r, z)$ gives a linear mass density distribution in the z-direction $\rho_{A,z}(r, z)$ (mass per unit length in the axial direction). In the special case where the plasma is only advancing in the z-direction, the values of r_{in} and r_{out} will be constant throughout the problem and the density will only be a function of z.

For the term in Eq. (4.4) representing mass accumulation through radial motion, we can integrate to yield

$$\int \int \int v_r \frac{\partial \rho_A}{\partial r} r \, dr \, d\theta \, dz = |v_r| \, A_2(r) \, \rho_A(r, z) = |v_r| \, \rho_{A,r}(r, z),$$
(4.6)

where A_2 is the surface area of a cylinder of radius r, which can be written for a cylinder of axial length l_z as $2\pi r l_z$. The density is again evaluated at the location (r, z) of the current sheet element for that moment in time. We take the magnitude of the radial velocity in this case because, while it is possible for the velocity to be negative (radially-inward motion) or positive (radially-outward motion), the mass accumulation due to motion in either direction will be positive. (Another way to explain this is that in the Lagrangian frame of motion, the components of velocity are always positive.) Combining A_2 and $\rho_A(r, z)$ gives a linear mass density distribution in the r-direction $\rho_{A,r}(r, z)$ (mass per unit length in the radial direction). In the special case where the plasma is only advancing in the r-direction, the density will only be a function of r and $\rho_{A,r}(r, z)$ will only have r-dependence through A_2.

Putting it all together, the equation for mass accumulation of a plasma moving in 2-D (as in Fig. 4.1b) is

$$\frac{dm_{\text{p}}}{dt} = \alpha \left[(|v_r| \, A_2(r) + v_z \, A_1(r_{\text{in}}, r_{\text{out}})) \, \rho_A(r, z) \right]. \quad (4.7)$$

The case in Fig. 4.1c for a plasmoid is slightly different. The plasmoid geometry as shown is inclusive of the centerline of the problem, so radial compression/collapse does not lead to additional mass accumulation in the plasma. Only mass entering the plasmoid through its axial motion needs to be included, but the cross sectional area A_1 presented to the gas changes as a function of time, and this change is dependent upon the velocity v_r, which governs the compression or expansion of the plasmoid. As such, the mass accumulation for this case is

$$\frac{dm_\mathrm{p}}{dt} = \alpha \left[v_z \, A_1 \left(r_\mathrm{out} \left(z \right) \right) \, \rho_A \left(z \right) \right]. \tag{4.8}$$

4.1.2 Momentum Conservation

4.1.2.1 Solid Bodies

The momentum equation is a statement of Newton's second law, which written for a solid conducting body of mass m results in

$$m\frac{d\mathbf{v}}{dt} = \sum \mathbf{F} \tag{4.9}$$

where the right-hand side represents the sum of the forces on the control volume. Solid bodies that do not collapse are only subject to the axial Lorentz body force given in Eq. (2.11). The sum of all pairwise electromagnetic interactions between external active circuits l through m and the moving body p yields a net axial momentum equation

$$m_\mathrm{p}\frac{dv_z}{dt} = \sum_{k=l}^{m} \frac{\partial M_{k\mathrm{p}}}{\partial z} I_k I_\mathrm{p}. \tag{4.10}$$

If the body is also collapsing, there will be some force $F_\mathrm{stiff}(t)$ associated with the stiffness of the body resisting collapse. Writing the form of this force, while beyond the scope of this monograph, requires a careful consideration of the structural dynamics of collapse when moving through the elastic and plastic deformation regimes (see, for example, Ref. [16]). Keeping the stiffness force in a general form, we can write the net radial momentum equation

$$m_\text{p} \frac{dv_r}{dt} = \frac{1}{2} \frac{\partial L_\text{p}}{\partial r} I_\text{p}^2 + \sum_{k=l}^{m} \frac{\partial M_{k\text{p}}}{\partial r} I_k I_\text{p} + F_\text{stiff}(t). \qquad (4.11)$$

In this equation, first term on the right-hand side will always be positive and resist radial motion of the body, the second term is negative and drives compression if the currents I_k and I_p are in opposing directions, and $F_\text{stiff}(t)$ resists compression.

4.1.2.2 Plasmas

The general equation for the momentum density $\rho \mathbf{v}$ of a plasma in the Lagrangian frame of reference is

$$\frac{\partial (\rho \mathbf{v})}{\partial t} + \mathbf{v} \cdot \nabla (\rho \mathbf{v}) = (\rho_i + \rho_e) \mathbf{E} + \mathbf{j} \times \mathbf{B} - \nabla p - \mathbf{f}_\text{res}, \quad (4.12)$$

where $\rho_e = -en_e$ is the local electron charge density, $\rho_i = Zen_i$ is the local ion charge density (with average charge Z), $\mathbf{j}$ is the local current density, p is the thermodynamic pressure, and $\mathbf{f}_\text{res}$ is the net effect of resistive forces. The resistive forces generally arise from collisions that exchange momentum and energy with particles outside the accelerated load. Moreover, because the power loss at extremely high velocities due to Bremsstrahlung radiation from Coulomb collisions scales as γ^6 (where γ is the relativistic Lorentz factor), radiation can also function as "internal braking" on the system. Therefore, adequate collisional-radiative models may be necessary to obtain a useful estimate of the resistive forces on the accelerated load.

A discussion of how to handle all the terms in Eq. (4.12) can be found in Ref. [17]. If we only consider length scales much larger than the Debye length (where the sum of the charge densities $\rho_e + \rho_i \approx 0$), and for the sake of brevity let the resistive forces go to zero, then this equation simplifies to

$$\frac{\partial (\rho \mathbf{v})}{\partial t} + \mathbf{v} \cdot \nabla (\rho \mathbf{v}) = \mathbf{j} \times \mathbf{B} - \nabla p. \qquad (4.13)$$

The first term on the right-hand side of the equation represents the electromagnetic Lorentz body force applied to the plasma. If we integrate this over the volume of the plasma, we obtain

$$m_p(t)\frac{dv_z}{dt} + \alpha\rho_{A,z}(r,z)v_z^2 = \sum_{k=l}^{m}\frac{\partial M_{kp}}{\partial z}I_kI_p - p_AA_{\text{cross}}\cos\theta,$$

$$m_p(t)\frac{dv_r}{dt} + \alpha\rho_{A,r}(r,z)v_r^2 = \frac{1}{2}\frac{\partial L_p}{\partial r}I_p^2 + \sum_{k=l}^{m}\frac{\partial M_{kp}}{\partial r}I_kI_p$$

$$- p_AA_{\text{cross}}\sin\theta, \qquad (4.14)$$

where the first term on the left in each equation represents acceleration of the plasma mass and the second term (with ρ_z and ρ_r indicating mass per unit length in the axial and radial directions, respectively) represents the momentum investment in entraining additional mass, which is accelerated from rest to the speed of the plasma. On the right-hand side we have replaced the Lorentz body force with the lumped-element circuit-derived forces from Eq. (2.11) and have written the sum of the pairwise electromagnetic interactions between external active circuits l through m and the moving plasma p. We have also included the action of the ambient gas pressure p_A exerting a force on the cross-sectional area perpendicular to the motion of the plasma A_{cross}. The angle θ is defined by the velocity components and equals

$$\theta = \tan^{-1}\left(\frac{v_r}{v_z}\right).$$

We can tailor Eqs. (4.14) for the plasmoid of Fig. 4.1. The second term in the axial equation still involves entrainment of gas in the manner described by Eq. (4.8). There will be no entrainment owing to radial motion of the plasma. However, the radial equation of motion is still important. It is critical that the location (r,z) of the moving lumped element being tracked through the integration of the momentum equation aligns with the radial and axial locations used in determining the self and mutual inductance functions $L_p(r)$ and

$M_{kp}(r, z)$, respectively. The ambient pressure term in the radial equation disappears, and the pressure term in the axial equation acts upon the area $A_1(r_{out}(z))$. This yields

$$m_p(t)\frac{dv_z}{dt} + \alpha A_1(r_{out}(z))\rho_A(z)v_z^2$$

$$= \sum_{k=l}^{m}\frac{\partial M_{kp}}{\partial z}I_kI_p - p_AA_1(r_{out}(z)),$$

$$m_p(t)\frac{dv_r}{dt} = \frac{1}{2}\frac{\partial L_p}{\partial r}I_p^2 + \sum_{k=l}^{m}\frac{\partial M_{kp}}{\partial r}I_kI_p, \qquad (4.15)$$

where the first equation governs the evolution of the axial velocity and the second governs the radial compression and expansion of the plasmoid.

In Section 3.2.1.1 we discussed a simpler version of Eq. (4.14) for a one-coil, one-body problem in one-dimension illustrated in Fig. 4.1a where it was assumed that $L_{C1} = L_p$, and that the total inductance $L_{tot}(z)$ and mutual inductance $\overline{M_{1p}}$ could be written in the form of Eqs. (3.15) and (3.16), respectively. Writing the impedances to the right of the dashed line in Fig. 3.3b as a Thévenin-equivalent impedance reduces the one-dimensional acceleration problem to that of a single current I_1 acting upon the inductor varying as $L_{tot}(z)$. Neglecting the effects of the ambient gas pressure, this yields a momentum equation

$$m_p(t)\frac{dv_z}{dt} + \rho_{A,z}(z)v_z^2 = \frac{1}{2}\frac{\partial L_{tot}}{\partial z}I_1^2 = \frac{\overline{L_{C1}}}{2z_0}\exp(-z/z_0)I_1^2.$$
$$(4.16)$$

4.1.3 Energy Conservation

One can stop at conservation of momentum, making assumptions about the pressure and conductivity of the moving conductor and how each relates to the temperature. However, if there is a desire to track the energy in a self-consistent manner, one must write a statement governing the time rate of change of energy in the moving conductor.

4.1.3.1 Solid Bodies

In general, a baseline estimate of the total energy of the accelerated load is:

$$E_{\text{tot}} = E_{\text{kin}} + E_{\text{therm}} + E_{\text{mag}}, \tag{4.17}$$

where the terms on the right side are kinetic, thermal, and magnetic field energy, respectively.

For a solid conductor, the conservation of energy equation for the conductor can be written as

$$\frac{dE_{\text{tot}}}{dt} = P_{\text{heat}} + P_{\text{work}} + P_{\text{mag}}. \tag{4.18}$$

The terms on the left-hand side are the time rate of change of E_{tot} and the terms on the right-hand side are the power in and out of the load associated with heating, electromagnetic work, and change in magnetic field energy, respectively.

All power deposited as magnetic field energy is stored in the field and does no work on the conductor. This essentially means

$$\frac{dE_{\text{mag}}}{dt} = P_{\text{mag}}.$$

Since the left-hand and right-hand sides of the equation are equal, we can drop them in the remainder of this section. (Note: this was more rigorously shown in Ref. [17].)

As a consequence, the problem reduces to one of properly modeling the power inputs on the right-hand side of Eq. (4.18) and determining how they are subdivided into the kinetic and thermal energy of the conductor.

For the sake of simplicity, assume that the solid, moving conductor is not collapsing ($v_r = 0$). The first term on the right-hand side of Eq. (4.18) is power deposited through Ohmic heating, which is

$$P_{\text{heat}} = I_{\text{p}}^2 R_{\text{p}}. \tag{4.19}$$

The term P_{work} is the electromagnetic work performed on the moving conductor by all other current-carrying conductors in the system. This can be found by taking the dot product of the velocity vector $\mathbf{v}$ and the electromagnetic force vector, the components of which are given in Eq. (2.11). For acceleration of a non-collapsing solid conductor acted upon by external active circuits l through m and using the definitions in Eq. (3.6), this becomes

$$P_{\text{work}} = \sum_{k=l}^{m} I_k I_{\text{p}} \frac{dM_{k\text{p}}}{dt}. \tag{4.20}$$

The time derivatives of Eq. (4.17) for a lumped-element moving solid conductor are

$$\frac{dE_{\text{tot}}}{dt} = m_{\text{p}} c_v \frac{dT}{dt} + \frac{d}{dt}\left(\frac{1}{2} m_{\text{p}} v_z^2\right), \tag{4.21}$$

where c_v and T in the first term on the right-hand side are the specific heat and temperature of the conductor and the second term is the change in kinetic energy. Substituting all the terms into Eq. (4.18) yields

$$m_{\text{p}} c_v \frac{dT}{dt} + \frac{d}{dt}\left(\frac{1}{2} m_{\text{p}} v_z^2\right) = I_{\text{p}}^2 R_{\text{p}} + \sum_{k=l}^{m} I_k I_{\text{p}} \frac{dM_{k\text{p}}}{dt}. \tag{4.22}$$

Finally, eliminating terms in the above equation using Eq. (4.10) yields

$$m_{\text{p}} c_v \frac{dT}{dt} = I_{\text{p}}^2 R_{\text{p}}, \tag{4.23}$$

which is an equation for the time-evolution of the temperature of the moving conductor. Here, we see explicitly how Ohmic heating of the conductor leads to an increase in temperature.

4.1.3.2 Plasmas

Following Ref. [5] we can write an equation governing the energy in a plasma moving through a gas as shown in Fig. 4.1b:

$$\frac{dE_{\text{tot}}}{dt} = P_{\text{heat}} + P_{\text{work}} + P_{\text{flux}} + P_{\text{A}} + P_{\text{rad}}, \tag{4.24}$$

where we have gained a term for power flux P_{flux} into the plasma owing to its motion through the ambient gas, a term P_A for power directed from the ambient gas into the plasma, and a radiative power P_{rad}. Each of these terms will be discussed in more detail below.

The total volume-integrated energy in the plasma is written as

$$E_{\text{tot}} = \frac{p}{\gamma - 1} \mathcal{V} + \frac{1}{2} m_p v^2, \qquad (4.25)$$

where p and γ are the pressure and ratio of specific heats of the plasma, respectively. Note that the pressure p represents the sum of the ion and electron pressures.

The Ohmic heating term P_{heat} is the same as Eq. (4.19). The electromagnetic work term is again found by taking the dot product of the electromagnetic force in Eq. (2.11) and the velocity vector. Using the definitions in Eq. (3.6) allows us to write

$$P_{\text{work}} = \frac{I_p^2}{2} \frac{dL_p}{dt} + \sum_{k=l}^{m} I_k I_p \frac{dM_{kp}}{dt}. \qquad (4.26)$$

The first term on the right-hand side represents work performed in changing the self-inductance of the conductor, which is accomplished by altering the geometric configuration of current in the conductor (primarily by collapse, compression, or expansion). The second term represents the electromagnetic work performed by the external circuits in accelerating the moving conductor.

The flux and entrainment of ambient gas into the plasma is a net power drain as that gas must be accelerated to the plasma speed. This can be written as

$$P_{\text{flux}} = -\alpha \left(\frac{1}{2} \rho_{A,z} (r, z) v_z^3 + \frac{1}{2} \rho_{A,r} (r, z) v_r^2 |v_r| \right). \qquad (4.27)$$

As illustrated by the last terms on the right-hand side of Eqs. (4.14), plasma moving into an ambient gas experiences a pressure that acts on the cross-sectional area perpendicular to the motion of the plasma. In addition, there will be a net flux of thermal energy into the plasma owing entrainment of gas. We can write these two effects into a combined term

$$P_A = \left(\frac{\alpha}{\gamma_A - 1} - 1\right) p_A A_{\text{cross}} |\mathbf{v}|, \tag{4.28}$$

where γ_A is the ratio of specific heats of the ambient gas. The positive first term in the parentheses corresponds to the entrained ambient gas thermal energy, while the negative term represents the ambient gas pressure force that resists plasma motion. Note that this formulation assumes negligible energy is transferred from the plasma to the ambient gas via thermal conduction and radiation.

Putting it all together, we have

$$
\begin{aligned}
\frac{dE_{\text{tot}}}{dt} &= I_{\text{p}}^2 R_{\text{p}} + \frac{I_{\text{p}}^2}{2} \frac{dL_{\text{p}}}{dt} + \sum_{k=l}^{m} I_k I_{\text{p}} \frac{dM_{k\text{p}}}{dt} \\
&\quad - \alpha \left[\frac{1}{2} \rho_{A,z}(r, z) v_z^3 + \frac{1}{2} \rho_{A,r}(r, z) v_r^2 |v_r| \right] \\
&\quad + \left(\frac{\alpha}{\gamma_A - 1} - 1\right) p_A A_{\text{cross}} |\mathbf{v}| + P_{\text{rad}}. \tag{4.29}
\end{aligned}
$$

The sum of all the terms on the right-hand side of Eq. (4.29) governs how the energy in Eq. (4.25) changes in time. The kinetic energy at any time is calculable through the results of the momentum equation, allowing Eq. (4.29) to be further simplified into the following form

$$\frac{d}{dt}\left(\frac{p\mathcal{V}}{\gamma - 1}\right) = I_{\text{p}}^2 R_{\text{p}} + \frac{\alpha}{\gamma_A - 1} p_A A_{\text{cross}} |\mathbf{v}| + P_{\text{rad}}. \tag{4.30}$$

The problem becomes one of determining the internal energy of the plasma, given as a function of the plasma pressure p and the ratio of specific heats γ. Terms on the right hand side of Eq. (4.30) represent contributions due to Ohmic heating of the plasma, entrainment of ambient gas internal energy, and radiation. Except for rare instances where plasma-heating via wave deposition is present, $P_{\mathrm{rad}} < 0$ due to line and Bremsstrahlung radiation.

Handling of Eq. (4.30) requires internal models and a closure approximation that connect the macroscopic plasma properties of pressure, temperature, and γ to the atomic processes in the plasma. This is discussed in greater detail in Section 5.3.

4.1.3.3 Additional Notes

A term that has not been explicitly defined at this point is the lumped-element resistance R_{p}, which is also dependent on the temperature. The variation of the resistivity of a metal as a function of temperature is typically a tabulated value. For a plasma the resistivity can be calculated through kinetic theory and is a function of temperature, both through the variation of the collision frequency with temperature and through the variation in the Coulomb collision cross-section with temperature and with ionization state, which is itself a function of temperature. Once the temperature is known and the resistivity calculated, it is a matter of integrating over the current distribution in the moving conductor to obtain the value of R_{p}. This was performed in Ref. [17].

One item not included in the formulation of the energy equation for a plasma is the initial ionization energy. This process occurs on a different timescale than the acceleration process, and in the present formulation the model assumes that there is a conducting medium already present when the external current pulse is initiated. Accurately modeling this initial ionization process within the framework of the model and quantifying the time-delay between external current pulse initiation and plasma formation is an area of ongoing research. This is also important in attempting to understand the effects of preionization (partial ionization of the working gas prior to the initiation of the external current pulse) on the plasma formation process.

Though the naturally high conductivity of solid metallic loads reduces the energy losses associated with propellant ionization (compared to *e.g.* a gaseous propellant), the high conductivity leads to low values of skin depth. This results in strong gradients in Ohmic heating between the surface of the metal and the interior, posing special challenges in estimating the time-dependent heating and thermal diffusion rates. These effects are compounded in a situation where multiple external circuits are all interacting with the accelerated body simultaneously and in a cyclical fashion.

4.2 MASS DISTRIBUTION AND ENTRAINMENT

Several different functional forms of ambient mass distribution have been explored in the literature. Recall that ambient mass distribution refers to the distribution of neutral gas mass just before a discharge event, and it is generally expressed in terms of a spatially-dependent density function. The losses associated with entrainment and acceleration of an ambient gas to the plasma velocity can vary greatly depending upon the distribution employed [18]. In this section, we explore a few of the more often used distributions.

4.2.1 Slug Mass

In this distribution, all the mass is initially contained in the moving conductor. This is the case for a solid conducting body, such as an imploding liner or coilgun projectile in vacuum. In this case, $\rho_A(r, z) = 0$ and there are no losses associated with mass entrainment. Also, since there is no gas in front of the moving body, there is no resistance to motion from the ambient gas pressure acting on the face of the conductive body.

4.2.2 Constant Mass Distribution

A constant ambient background gas density $\rho_A(r, z) =$ constant is relatively straightforward to implement. However, it does result in high losses since the conducting plasma encounters significant amounts of gas and expends power to accelerate that gas to the plasma velocity even as the sustaining electromagnetic force applied by the external coil(s) is decreasing with increasing separation distance.

4.2.3 Linear Mass Distribution

In Refs. [11, 12] a linear mass density distribution was employed for a flat planar coil axial plasma accelerator. The injected mass produced a distribution that was highest at the coil face and decreased linearly with distance to a depth δ_m. This is written as

$$\rho_{A,z}(z) = \begin{cases} \rho_0 A_1\left(r_{\text{in}}, r_{\text{out}}\right)\left(1 - z/\delta_m\right), & z \leq \delta_m, \\ 0, & z > \delta_m, \end{cases} \quad (4.31)$$

where ρ_0 is the value of the density at the coil face. Most of the gas is entrained while the plasma is still close to the coil face of the external circuit. Consequently, while the entrainment losses are greater than those of the slug mass loading, they are much lower than the constant mass distribution since the losses are a function of the square of the plasma velocity. This allows the plasma moving through a linear mass distribution to entrain most of the gas at lower velocities and undergo significant electromagnetic acceleration after most of the gas has been entrained.

4.3 TRANSITION FROM RADIAL TO AXIAL MOTION

In the literature, most notably for pulsed plasma thrusters, there exists a significant body of work contemplating the mechanism by which radial motion of a plasma is somehow "converted" into directed axial motion. For thruster applications, this is important because only plasma propellant leaving the thruster control volume in the axial direction generates thrust.

In this section, we say a few words on the mechanism that drives this transition. We shall show that the two-dimensional momentum equation formulation for lumped-element acceleration previously presented in this section self-consistently handles this transition of plasma motion from the radial to the axial direction.

We use as our example the case of two-dimensional inductive acceleration that occurs in a conical theta pinch, where the motion is initially in both the radial and axial directions, but as the plasma collapses towards the centerline the radial motion will eventually halt leaving only axial motion. Two aspects of this inductive acceleration process must be captured in a model:

1. Ensuring the conservation of the momentum vector in the radial and axial directions and

2. Accounting for the mechanisms that cause the change in the direction of the plasma motion from radial to axial.

The conservation of the momentum of this simple system can be tracked by considering a system consisting of a single external coil 1 and a plasma p, where the current in coil 1, I_1, creates a magnetic field B_1 and has a typical current waveform as shown in Fig. 4.2. $I_1 > 0$ (current density vector $\mathbf{j}_1$ in the $+\hat{\theta}$ direction) between $0 < t < \tau_{\text{rise}}$ (region shaded in dark gray in Fig. 4.2). During the rise, $dI_1/dt > 0$ and $dB_1/dt > 0$ through Ampère's law. This induces an opposing electric field in the plasma ($\mathbf{E}$ in the $-\hat{\theta}$ direction) through Faraday's law. As a result, the induced voltage in the plasma is

$$V_{\text{p}} = -\int \mathbf{E} \cdot d\mathbf{l} > 0,$$

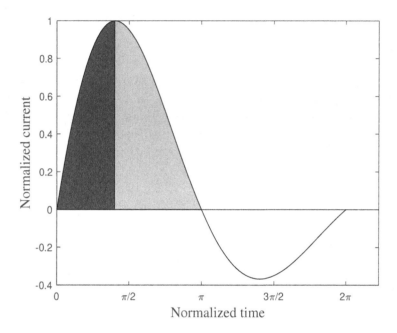

Figure 4.2 Schematic of a current trace in a conical theta pinch, with $\tau_{\text{period}} = 2\pi$ with rising (dark gray region) and decreasing (light gray region) current. Note that $\tau_{\text{rise}} < \pi/2$ for an inductively-coupled accelerator.

between $0 < t < \tau_{\text{rise}}$, where the path integral is performed in the direction of the azimuthal plasma current $d\mathbf{l}$.

As the coil current decreases between $\tau_{\text{rise}} < t < \tau_{\text{period}}/2$ (region shaded in light gray in Fig. 4.2), $dI_1/dt < 0$ and $dB_1/dt < 0$. This induces an opposing electric field in the plasma ($\mathbf{E}$ now in the $+\hat{\theta}$ direction). The induced voltage in the plasma between $\tau_{\text{rise}} < t < \tau_{\text{period}}/2$ is

$$V_{\text{p}} = -\int \mathbf{E} \cdot d\mathbf{l} < 0.$$

A circuit model that produces this self-consistent relationship between the time-varying current in the coil and the induced voltage in the plasma will also ensure conservation of the momentum vector.

Consider the 1-D analogous mechanical model of two blocks attached by an intermediate compressed spring. The stored potential energy from the expanding spring (EM field here) becomes the kinetic energy of the moving blocks; the spring imparts equal and opposite forces to the block on the left (coil here) and the right (plasma here). Our focus here is on the interaction between the expanding spring and the block on the right. As the motion of the plasma transitions from the radial to the axial direction, the force exerted by the electromagnetic field changes the momentum of the plasma ($\mathbf{P}$) with time,

$$\mathbf{F}_{\text{field}} = \frac{d\mathbf{P}_{\text{plasma}}}{dt}.$$

This can be re-written per unit volume as

$$\mathbf{f}_{\text{field}} = \frac{d\mathbf{p}_{\text{plasma}}}{dt}. \tag{4.32}$$

The force per volume on the left hand side can be rewritten [19] using the Ampère-Maxwell equation, the Maxwell stress tensor, $\bar{\bar{\mathsf{M}}}_B$, and the Poynting flux, $\mathbf{S} = \mathbf{E} \times \mathbf{B}/\mu_0$ as

$$\mathbf{f}_{\text{field}} = \mathbf{j} \times \mathbf{B} = \nabla \cdot \bar{\bar{\mathsf{M}}}_B - \epsilon_0 \mu_0 \frac{\partial \mathbf{S}}{\partial t}. \tag{4.33}$$

Thus, the force per volume has a spatially varying component and a time-varying component with the former being the dominant term. Combining Eqs. (4.33-4.32), the result is

$$\nabla \cdot \bar{\bar{\mathsf{M}}}_B - \epsilon_0 \frac{\partial \left(\mathbf{E} \times \mathbf{B}\right)}{\partial t} = \frac{\partial \mathbf{p}_{\text{plasma}}}{\partial t}. \tag{4.34}$$

Note that the second term on the left hand side is the time derivative of the momentum per volume in the electromagnetic fields ($\mathbf{p}_{\text{field}} = \epsilon_0 \mathbf{E} \times \mathbf{B}$), which consists of contributions from the electromagnetic fields created by the coil current I_1 and inductively-driven current in the plasma. The right hand side is the time derivative of the momentum per volume of the plasma ($\mathbf{p}_{\text{plasma}} = \rho \mathbf{v}$). Because momentum is conserved individually in both the radial and axial directions,

$$\left(\nabla \cdot \bar{\bar{\mathsf{M}}}_B\right)_r - \frac{\partial p_{\text{field},r}}{\partial t} = \frac{\partial p_{\text{plasma},r}}{\partial t},$$

$$\left(\nabla \cdot \bar{\bar{\mathsf{M}}}_B\right)_z - \frac{\partial p_{\text{field},z}}{\partial t} = \frac{\partial p_{\text{plasma},z}}{\partial t}.$$

If $\mathbf{E} = E_\theta \, \hat{\theta}$ and $\mathbf{B} = B_r \, \hat{r} + B_z \, \hat{z}$, then

$$\left(\nabla \cdot \bar{\bar{\mathsf{M}}}_B\right)_r - \frac{\partial \left(\epsilon_0 E_\theta B_z\right)}{\partial t} = \frac{\partial \left(\rho v_r\right)}{\partial t}, \tag{4.35a}$$

$$\left(\nabla \cdot \bar{\bar{\mathsf{M}}}_B\right)_z - \frac{\partial \left(-\epsilon_0 E_\theta B_r\right)}{\partial t} = \frac{\partial \left(\rho v_z\right)}{\partial t}. \tag{4.35b}$$

The initial plasma acceleration in a conical theta pinch is radially inward and axially forward. To understand the conversion of plasma momentum from the radial to the axial direction, one must account for the opposite change in the electromagnetic momentum. Plasma acceleration in the radially-inward (negative) direction is consistent with $d(E_\theta B_z)/dt > 0$ and axial acceleration in the forward (positive) direction aligns with $d(-E_\theta B_r)/dt < 0$.

The values B_r and B_z represent components of the same solenoidal magnetic field produced by a fixed external coil and inductively-driven currents is the plasma. Consequently, the signs of dB_r/dt and dB_z/dt are the same as that of dB_1/dt. Thus, $d(E_\theta B_{1z})/dt > 0$ also implies $-d(E_\theta B_{1r})/dt < 0$ for any E_θ in the plasma, ensuring that an increase in the electromagnetic radial momentum also produces a decrease in the electromagnetic axial momentum. Therefore, due to the coupling between the electromagnetic momentum and the plasma momentum in Eqs. (4.35), a decrease in the plasma radial momentum is consistent with an increase in the plasma axial momentum.

The dominant term in the left hand side of Eq. (4.34) is the spatial variation of the magnetic field and, therefore,

Eq. (4.11) emphasizes only that part of it. In the lumped-element circuit model formulation, this self-consistent conversion process manifests in the inductance of the circuit-plasma system. In any inductively-coupled pulsed accelerator where the accelerated body is capable of motion in the r and z directions, the force on that body is given by Eq. (2.11), rewritten here in vector form as for a single coil as

$$\mathbf{F}(r, z) = \frac{I_p^2}{2}\nabla L_p + I_1 I_p \nabla M_{1p}. \tag{4.36}$$

Note that the coil inductance term does not appear in the force equation. This is because the coil's inductance is only a function of the path of the current flowing through it (and the permeability of the surroundings). Therefore there is no impact on the coil inductance when the plasma current moves away from the coil, even though there is an impact on the inductance presented to the coil's driving circuit. If the current path through the plasma were to not change shape throughout the displacement of the plasma current from the coil then the plasma inductance term would also be zero, leaving only the changing mutual inductance term as the driving force on the plasma current as it moves away from the coil.

Physically, the first term on the right-hand side containing the gradient in the plasma self inductance resists compression while the second term with the gradient in mutual inductance leads to both compression and axial acceleration of the body. As the body moves in the axial direction, the time-evolution of a curl-free magnetic field results in a mutual inductance gradient that is more axial than radial. In addition, we observe from the example conical theta pinch mutual inductance axial profiles shown in Fig. 3.4b that the axial gradient in mutual inductance is much greater if the body has a greater radius. The resistance to compression provided by the self inductance term results in acceleration occurring at a greater radial position than if the radial compression was unrestrained. Accurate representations of the self inductance L_p

as function of r and the mutual inductance M_{1p} as a function of r and z will capture all these effects related to converting radial motion to axial acceleration in an inductively-coupled pulsed accelerator.

CHAPTER 5

Internal Models

M ODELING the pulsed-inductive acceleration process in terms of circuit parameters such as resistance and inductance requires lumped-sum estimates of internal properties of the accelerating field as well as the accelerated load. Specifically, those estimates are necessary because circuit values (like R) represent volume integrals of spatially varying quantities (such as resistivity, η) and those differential quantities, in turn, can depend upon other local values (like density and temperature). This chapter provides the link between the "microscopic" physical processes that drive the behavior of the system and the "macroscopic" circuit model view that we use in this work to describe the system, estimating the circuit parameters by appropriately accounting for relevant internal processes.

5.1 LUMPED-CIRCUIT METHOD

There are three key aspects to the approach connecting the internal mechanisms of the acceleration process to the externally-observed behavior:

1. Start with the underlying physics in a differential element and integrate to obtain the volume-averaged values of relevant properties that will be useful in a circuit model.

DOI: 10.1201/9780429351976-5

2. Ensure consistency between the microscopic phenomena within the differential element and the resulting macroscopic properties calculated through volume integration over the domain.

3. Identify zeroth-order constant quantities and first-order change approximations.

Following item #1, when a differential element of an arbitrary quantity varies in space and time, $f(x, y, z, t)$, integrating this function over a volume (e.g., the volume of the accelerated body) transforms it as:

$$\int_{\mathcal{V}} f(x, y, z, t) \, d\mathcal{V} = F(t). \tag{5.1}$$

Any partial derivative with respect to time of f in a physical model will be transformed into an ordinary time derivative of F in the circuit model. Since one of the benefits of a circuit model is the ability to analyze the acceleration process as only a function of time, it is valuable to use the volume integral approach to simplify the problem by removing the spatial dependence through aggregation into lumped-sum values.

The main emphasis of this chapter is on item #2, ensuring that the physical implications of the differential quantities, such as mass density ρ, current density j, electric field E, magnetic field B, and resistivity η are preserved and appropriately translated into volume-integrated lumped-sum quantities such as the mass of the accelerated load m, net current in a conductor I, voltage V, inductance L, resistance R, and so forth.

As implied by item #3, a lumped-sum model represents an average that removes certain small-scale phenomena occurring within the volume of the accelerated load. Let us consider some noteworthy phenomena that are hidden when calculating lumped-sum quantities for the circuit model.

5.1.1 Instabilities

Acceleration of a plasma can give rise to local instabilities of the Kelvin-Helmholtz (KH) and the Rayleigh-Taylor (RT) types. Generally, such local instabilities can be detrimental to the desired macroscopic behavior of the system and can hinder a clean transition from the differential description to an integral lumped-element circuit description of a plasma. As noted by Polzin *et al.* [1], "the development of large scale instabilities could lead to plasma filamentation, reducing the inductive coupling and making the current sheet more permeable to both encountered gas and the magnetic field driving acceleration."

In particular, it is important to consider the presence of RT instabilities in inductive accelerators because the acceleration process often involves magnetic pressure of a lower-density region pushing against a higher-density plasma. The growth rate of RT instabilities, γ, can be calculated using [20],

$$
\gamma = \sqrt{\left\{ a_g \, k \left(\frac{\rho_{\text{high}} - \rho_{\text{low}}}{\rho_{\text{high}} + \rho_{\text{low}}} \right) \right\} - \left\{ \frac{2 \left(\mathbf{k} \cdot \mathbf{B}_{\text{app}} \right)^2}{\mu_0 \left(\rho_{\text{high}} + \rho_{\text{low}} \right)} \right\}},
$$

$$(5.2)$$

where the region of high density ρ_{high} is pushed by a region of low density ρ_{low}, with an acceleration a_g that effectively acts in the direction opposing motion. The resulting RT oscillation has a vector wave number $\mathbf{k}$ with magnitude k. While the terms inside the first set of brackets $\{\dots\}$ cause the growth of the instability, the terms inside the second set of brackets $\{\dots\}$ cause the instability to decay. Figure 5.1 shows the growth rate of an instability of wavelength $\lambda = 1$ cm (where $\lambda = 1/k$) for a range of applied magnetic field values and high-to-low density ratios ρ_{ratio}. In this case, the magnetic field $\mathbf{B}_{\text{app}}$ is applied parallel to $\mathbf{k}$. These data illustrate how a magnetic field used in this manner can diminish the growth rate of RT instabilities. For the case of an inductively-coupled pulsed accelerator, if the growth rate γ is slow compared to the time scale of acceleration τ_a, such that $\gamma\, \tau_a \ll 1$, then this local phenomenon will not result in observable macroscopic effects or significantly impact the acceleration process.

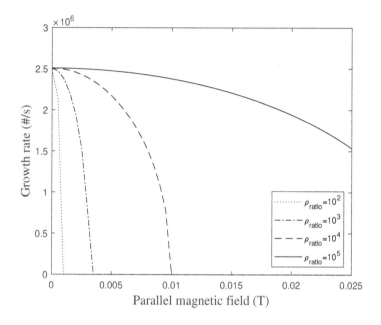

Figure 5.1 Growth rate of a Rayleigh-Taylor instability as a function of a magnetic field applied parallel to the direction of the propagation vector **k** for various high-to-low density ratios ρ_{ratio}. ($a_g = 10^{10}$ m/s^2, $\lambda = 1$ cm, $\rho_{\text{low}} = 10^{-9}$ kg/m^3.)

Metallic solid liners accelerated (imploded) by pulsed electromagnetic fields can experience buckling. While this represents an instability from a structural mechanics perspective, buckling may also have a beneficial effect in stiffening the material to prevent radial distortion. It may even be structurally advantageous to intentionally design liners with "pleats" to reduce distortions during compression [21]. Furthermore, a background applied magnetic field can also serve to reduce radial distortion during the compression of the liner [16].

Despite the reality of instabilities in both solid metallic conductors and plasmas, accelerator designs can incorporate suitable techniques that suppress instability growth rates [21] or shorten the acceleration timescale to prevent small-scale instabilities from affecting accelerator performance. Consequently, to zeroth-order, such local instabilities do not affect the results obtained using a lumped-element circuit model of the acceleration process.

5.1.2   Skin Effect

Unlike in conductors carrying steady direct currents, the charge carriers in situations where the current is time-varying are not uniformly distributed throughout the conductor but are instead concentrated near its surface. A time-dependent current begins at the surface and diffuses into the material until the current reverses direction. This phenomenon where the current flows mainly through the outer layer of the conductor is known as the skin effect, and it is related to the ability of a conductor to shield against diffusion of the magnetic field into a highly-conductive medium.

The diffusion equation describes this process by which the current distribution is established in a conductor as a function of time. The solution yields a simple solution of the form

$$\mathbf{j}(\delta) = \mathbf{j}_{\text{surf}} \exp(-\delta/\delta_{\text{skin}}), \tag{5.3}$$

where $\mathbf{j}(\delta)$ is the current density as a function of penetration depth δ perpendicular to the conductor surface and $\mathbf{j}_{\text{surf}}$ is the current density at the surface. The skin depth $\delta_{\text{skin}} \simeq \sqrt{\eta/(\mu_0 \omega)}$ is the characteristic diffusion length scale, where μ_0 is the magnetic permeability of free space, ω is the angular frequency of the current, and η is the resistivity of the conductor.

The exponential decay of current density from the surface into the bulk of the conductor has implications for the volume integral. Furthermore, because the energy deposition due to Ohmic heating varies as the square of the current density, the skin effect produces a strong gradient in the energy deposition profile. In the previous chapter it was noted that changes in the distribution of current in each conductor can introduce variations in the self and mutual inductance values that one obtains when performing volume integrals to obtain these lumped-element parameters. While we have not done so in this work, it is possible in principle to account for these effects depending upon the level of control a user has in fixing the current distribution in the magnetic field solver used to calculate inductance values.

5.2 STATE OF MATTER

The state of matter of the accelerated body affects its inductive response to a changing externally-imposed field and, consequently, to a circuit model of that process. Two states of matter that have adequate electrical conductivity to support the types of inductively-coupled pulsed acceleration considered in this work are discussed in this section.

5.2.1 Metallic Solids

When a metallic solid is inductively accelerated, as in a coil-gun [6, 22], there are some internal phenomena that are relevant to the formulation of the circuit model.

1. Some of the power transferred from the circuit to the load is directed into heating, deformation, and melting of the conductor and not into the acceleration process. Furthermore, as the temperature of the accelerated body increases, the already strong spatial gradient in the current profile arising due to the skin effect described in Eq. (5.3) can further increase owing to resistivity being a function of temperature. Because the timescale of the acceleration is typically faster than the timescale to reach thermal equilibrium, the resistivity will not reach a uniform value during an acceleration event, or pulse. Consequently, the value of resistance

and the resistive power dissipation in the circuit model are only volume-averaged quantities.

2. The skin effect is more pronounced in metallic solids owing to the low value of resistivity in comparison to a plasma. This is not necessarily a detriment and is, in fact, an inherent feature of a coilgun, where the accelerated body is designed to be thick enough to act as a flux-excluding shell that minimizes Ohmic heating inside the bulk of the volume [23].

3. In addition to confining the electromagnetic force to the skin, the strong non-uniformity in current density (c.f. Eq. (5.3)) also affects the calculation of mutual inductance.

5.2.2 Plasmas

There are similarities and differences in the thermodynamics and electromagnetic processes that affect inductively-accelerated plasmas when compared to metallic solids.

1. Unlike metallic solid accelerators, plasma accelerators require significant energy transfer from the external circuit into the load to first ionize the working fluid before it will support an inductively-driven current. The fraction of the input power directed into ionization and even the fraction of the gas that is ionized are in general not a constant with respect to time and both are dependent upon the comparative timescales for gas ionization and acceleration, the type of gas used, and whether any part of the gas has been preionized (ionized prior to initiation of the acceleration discharge in the circuit).

2. After creation of the plasma, some of the power from the external circuit is deposited into further heating the gas and increasing its internal energy, as in a metallic solid. The partition function for the plasma governs the

ionization state of the plasma and the electronic excitation levels populated by bound electrons in the ions. The gas could also be in a non-equilibrium ionziation state at any point in the discharge, further complicating matters. Energy directed into heating the gas or increasing the internal energy of the ions is generally not available for acceleration, limiting the overall efficiency in converting initially-stored energy into directed kinetic energy of the exhaust [17].

3. The resistivity is much greater in a plasma than in a metallic solid (by several orders of magnitude). While this results in the skin depth being larger in a plasma, it is still a significant effect resulting in spatial non-uniformities in the Ohmic heating profile and the application of the Lorentz body force to the plasma. As with the metallic solid, this non-uniform current distribution can also alter the mutual inductance between the plasma and the external circuit.

5.3 EQUATIONS-OF-STATE AND CLOSURE

Due to the strong influence of the state of the matter on the transfer of energy and momentum from an external driving circuit into an inductively-accelerated load, it is necessary to have equations-of-state (EOS) to reliably model the internal conditions of the latter. Writing a set of equations that govern the electrodynamics, thermodynamics, and hydrodynamics of the system results in a set of relationships with more unknowns than equations. Thus, equations of state provide the necessary closure to the system by relating several unknown variables to each other and increasing the number of equations to equal the number of unknowns.

Modeling the internal state of accelerated/collapsing metallic liners may require appropriate relationships between the external (electromagnetic) pressure and the volume of the solid. General continuum models, such as the Birch-Murnaghan EOS [24], the Mie-Grüneisen EOS [25], or other expressions specifically suited for these conditions are required to provide closure.

For inductively-accelerated plasmas, the thermodynamic pressure and the ratio of specific heats must be related to the density and temperature as $p(\rho, T)$ and $\gamma(\rho, T)$, respectively. For an accelerated plasma in thermal equilibrium, the calculation of thermodynamic pressure could require the knowledge of the variation of the total partition function with volume $Q(V)$, from which one can obtain,

$$p = Nk_BT\frac{\partial \ln Q}{\partial V}, \tag{5.4}$$

where N is the number of particles and k_B is Boltzmann's constant. We note that estimating $Q(V)$ for interacting particles is a non-trivial task and it is the controlling factor in determining how much the plasma behavior deviates from an ideal gas EOS.

Knowledge of the pressure distribution in the domain can be used to compute the net force on the accelerated plasma per unit volume that arises due to a pressure gradient $(-\nabla p)$ and the thermal energy per unit volume of plasma $(p/(\gamma-1))$, which includes contributions from both ions and electrons. The high temperatures associated with pulsed plasmas can make internal energy modes in the ions (such as electronic excitation) available as intermediate sinks, altering the value of γ as a function of temperature. A rigorous estimation of γ requires a calculation of the plasma specific heats at constant pressure c_p and constant volume c_v in this set of steps,

$$\mathcal{E} = \frac{N}{V}k_BT^2\frac{\partial \ln Q}{\partial T}, \tag{5.5}$$

$$h = \mathcal{E} + p, \tag{5.6}$$

$$c_p = \frac{\partial h}{\partial T}, \tag{5.7}$$

$$c_v = \frac{\partial \mathcal{E}}{\partial T}, \tag{5.8}$$

$$\gamma = \frac{c_p}{c_v}, \tag{5.9}$$

where $\mathcal{E}$ is the internal energy of the ions in the gas and h is the enthalpy. Though cumbersome, it is possible to incorporate these internals models into a circuit model of pulsed inductive acceleration, as shown in Ref. [17]. To reduce computational complexity, a first-order estimate could assume a single-fluid approximation and treat ions and electrons together and employ an ideal gas model for $p(\rho, T)$ to obtain a less accurate but suitable value for γ.

CHAPTER 6

Illustrative Results

I N this monograph, we have spent considerable time
discussing the means of modeling different types of
inductively-coupled pulsed accelerators using a lumped-
element circuit model approach. These modeling techniques
have been employed in investigations of different accelerator
configurations, and in this section we present an overview of
the results from a select number of those studies.

6.1 SETS OF MODELING EQUATIONS

Before diving into a presentation of the data found in the literature, it is perhaps worthwhile to consolidate in this one place
complete sets of modeling equations for different accelerator types. In general, we shall see that, even neglecting some
terms and equations previously derived, the lumped-element
circuit modeling method represents a set of tightly-coupled
equations that can become quite expansive going from one-
dimensional to two-dimensional acceleration and modeling
multiple external circuits.

6.1.1 Solid Non-Collapsing Projectile

The solid non-collapsing projectile, or macron, is the basis for
the coilgun type accelerator. It moves only in the z-direction,
and when it is acted upon by multiple external coils the

DOI: 10.1201/9780429351976-6

following set of governing equations result:

$$V_{Ci} = R_{ei}I_i + (L_{0i} + L_{Ci})\frac{dI_i}{dt}$$

$$+ \frac{d}{dt}(M_{ip}I_p) + \sum_{\substack{j=l \\ j \neq i}}^{m} M_{ij}\frac{dI_j}{dt}, \quad \forall\, i = [l, m], \quad (6.1a)$$

$$0 = R_p I_p + L_p\frac{dI_p}{dt} + \sum_{k=l}^{m} \frac{d}{dt}(M_{kp}I_k), \quad (6.1b)$$

$$\frac{dV_{Ci}}{dt} = -\frac{I_i}{C_i}, \quad \forall\, i = [l, m], \quad (6.1c)$$

$$\frac{dM_{kp}}{dt} = \frac{\partial M_{kp}}{\partial z}v_z, \quad \forall\, k = [l, m], \quad (6.1d)$$

$$\frac{dz}{dt} = v_z, \quad (6.1e)$$

$$m_p\frac{dv_z}{dt} = \sum_{k=l}^{m} \frac{\partial M_{kp}}{\partial z}I_k I_p. \quad (6.1f)$$

In this set, there are separate Eqs. (6.1a), (6.1d), (6.1d) for every individual external conductor l through m.

6.1.2 Axially-Accelerating Plasma

Many studies using the lumped-element circuit model have been performed on planar inductive pulsed plasma thrusters, which act upon and accelerate propellant in the z-direction. Typically in these thrusters there is only one coil acting upon the plasma disk. If we neglect the ambient pressure forces acting on the disk, we obtain the following set of equations:

$$V_{C1} = R_{e1}I_1 + (L_{01} + L_{C1})\frac{dI_1}{dt} + \frac{d}{dt}(M_{1p}I_p), \quad (6.2a)$$

$$0 = R_p I_p + L_p\frac{dI_p}{dt} + \frac{d}{dt}(M_{1p}I_1), \quad (6.2b)$$

$$\frac{dV_{C1}}{dt} = -\frac{I_1}{C_1}, \quad (6.2c)$$

$$\frac{dM_{1p}}{dt} = \frac{\partial M_{1p}}{\partial z}v_z, \quad (6.2d)$$

$$\frac{dz}{dt} = v_z, \tag{6.2e}$$

$$m_p \frac{dv_z}{dt} = \frac{\partial M_{1p}}{\partial z} I_1 I_p - \alpha \rho_{A,z}(z) v_z^2, \tag{6.2f}$$

$$\frac{dm_p}{dt} = v_z \, \alpha \rho_{A,z}(z), \tag{6.2g}$$

where $\rho_{A,z}$ is the ambient gas mass per unit length in the axial direction. Most studies have not employed the energy equation formulation, but it can be included in the manner described in Ref. [17].

6.1.2.1 Simplifications of Section 3.2.1.1

In Section, 3.2.1.1, it was noted that additional simplifications have been made in past works where an axially accelerating plasma disk was modeled. Those simplifications and the observation that the terminal inductance could be fit to an analytical function of z resulted in, among other things, the writing of the mutual inductance in a functional form, which could be differentiated to yield statements for the spatial gradient and time rate of change in that term. Those assumptions resulted in the following simplified set of equations:

$$V_{C1} = R_{e1} I_1 + \left(L_{01} + L_{th}|_{z=0} + \overline{L_{C1}} \right) \frac{dI_1}{dt}$$
$$+ \frac{d}{dt} \left(\overline{M_{1p}} I_p \right), \tag{6.3a}$$

$$0 = R_p I_p + \overline{L_{C1}} \frac{dI_p}{dt} + \frac{d}{dt} \left(\overline{M_{1p}} I_1 \right), \tag{6.3b}$$

$$\frac{dV_{C1}}{dt} = -\frac{I_1}{C_1}, \tag{6.3c}$$

$$\frac{d \left(\overline{M_{1p}} \right)}{dt} = -\frac{\overline{L_{C1}}}{2z_0} \exp\left(-z / (2z_0) \right) v_z, \tag{6.3d}$$

$$\frac{dz}{dt} = v_z, \tag{6.3e}$$

$$m_p \frac{dv_z}{dt} = \frac{\overline{L_{C1}}}{2z_0} \exp\left(-z/z_0 \right) I_1^2 - \alpha \rho_{A,z}(z) v_z^2, \tag{6.3f}$$

$$\frac{dm_p}{dt} = v_z \, \alpha \rho_{A,z}(z). \tag{6.3g}$$

6.1.3 Conical Theta-Pinch Plasma

There have been a limited number of lumped-element circuit model studies on conical theta-pinch plasmas, where the plasma moves in the r-z plane. If we neglect the ambient pressure forces acting on the plasma and do not include the energy equation formulation, we obtain for a system with multiple external circuits:

$$V_{Ci} = R_{ei}I_i + (L_{0i} + L_{Ci})\frac{dI_i}{dt}$$

$$+ \frac{d}{dt}(M_{ip}I_p) + \sum_{\substack{j=l \\ j\neq i}}^{m} M_{ij}\frac{dI_j}{dt}, \qquad \forall\, i = [l, m], \tag{6.4a}$$

$$0 = R_p I_p + \frac{d}{dt}(L_p I_p) + \sum_{k=l}^{m} \frac{d}{dt}(M_{kp}I_k), \tag{6.4b}$$

$$\frac{dV_{Ci}}{dt} = -\frac{I_i}{C_i}, \qquad \forall\, i = [l, m], \tag{6.4c}$$

$$\frac{dL_p}{dt} = \frac{\partial L_p}{\partial r}v_r, \tag{6.4d}$$

$$\frac{dM_{kp}}{dt} = \frac{\partial M_{kp}}{\partial r}v_r + \frac{\partial M_{kp}}{\partial z}v_z, \qquad \forall\, k = [l, m], \tag{6.4e}$$

$$\frac{dz}{dt} = v_z, \tag{6.4f}$$

$$\frac{dr}{dt} = v_r, \tag{6.4g}$$

$$m_p\frac{dv_z}{dt} = \sum_{k=l}^{m} \frac{\partial M_{kp}}{\partial z}I_k I_p - \alpha\rho_{A,z}(r, z)v_z^2 \tag{6.4h}$$

$$m_p\frac{dv_r}{dt} = \frac{1}{2}\frac{\partial L_p}{\partial r}I_p^2 + \sum_{k=l}^{m} \frac{\partial M_{kp}}{\partial r}I_k I_p - \alpha\rho_{A,r}(r, z)v_r^2, \tag{6.4i}$$

$$\frac{dm_p}{dt} = \alpha\left[(|v_r|\,A_2(r) + v_z\,A_1(r_{in}, r_{out}))\,\rho_A(r, z)\right], \tag{6.4j}$$

where $\rho(r, z)$ is the ambient gas density and ρ_z and ρ_r are the mass density multiplied by A_1 and A_2, respectively.

6.1.4   Plasmoid Accelerator

There have been even fewer lumped-element circuit model studies on plasmoid accelerators, though this technique appears to be gaining in popularity. Again neglecting the ambient pressure forces acting on the plasmoid and not including the energy equation, we obtain:

$$V_{Ci} = R_{ei}I_i + (L_{0i} + L_{Ci})\frac{dI_i}{dt}$$

$$+ \frac{d}{dt}(M_{ip}I_p) + \sum_{\substack{j=l \\ j\neq i}}^{m} M_{ij}\frac{dI_j}{dt}, \qquad \forall\, i = [l, m],$$

$$(6.5a)$$

$$0 = R_p I_p + \frac{d}{dt}(L_p I_p) + \sum_{k=l}^{m}\frac{d}{dt}(M_{kp}I_k), \qquad (6.5b)$$

$$\frac{dV_{Ci}}{dt} = -\frac{I_i}{C_i}, \qquad \forall\, i = [l, m], \qquad (6.5c)$$

$$\frac{dL_p}{dt} = \frac{\partial L_p}{\partial r}v_r, \qquad (6.5d)$$

$$\frac{dM_{kp}}{dt} = \frac{\partial M_{kp}}{\partial r}v_r + \frac{\partial M_{kp}}{\partial z}v_z, \qquad \forall\, k = [l, m], \qquad (6.5e)$$

$$\frac{dz}{dt} = v_z, \qquad (6.5f)$$

$$\frac{dr}{dt} = v_r, \qquad (6.5g)$$

$$m_p\frac{dv_z}{dt} = \sum_{k=l}^{m}\frac{\partial M_{kp}}{\partial z}I_k I_p - \alpha A_1\left(r_{out}\left(z\right)\right)\rho_A\left(z\right)v_z^2, \qquad (6.5h)$$

$$m_p\frac{dv_r}{dt} = \frac{1}{2}\frac{\partial L_p}{\partial r}I_p^2 + \sum_{k=l}^{m}\frac{\partial M_{kp}}{\partial r}I_k I_p, \qquad (6.5i)$$

$$\frac{dm_p}{dt} = \alpha\left[v_z\, A_1\left(r_{out}\left(z\right)\right)\rho_A\left(z\right)\right]. \qquad (6.5j)$$

Note that the plasmoid model is identical to the conical theta-pinch model with the exception that mass entrainment due to radial compression/expansion no longer appears in the mass conservation and radial momentum equations.

6.2   SELECT RESULTS AND EXAMPLES

We proceed with a presentation of select modeling results from various studies, which illustrate the utility of the modeling technique in studying various applications.

6.2.1   Comparison of Models from Section 6.1.2

Two models of axial plasma acceleration were presented in Section 6.1.2. Equation set (6.2) represents a general-form equation set, while equation set (6.3) employed additional assumptions in the derivation. The latter equation set represents the one that has been employed more often to study planar inductive pulsed plasma thrusters, so it is reasonable to ask how well the results of the two models compare.

Using the inductance profiles presented in Fig. 3.6, specifically that found using Eq. (3.11) for the general-form equation set and those representing Eqs. (3.15) and (3.16) for the simplified equation set, we can solve the sets of coupled first-order ordinary differential equations numerically. Solutions were generated for the parameters given in Table 6.1.

Table 6.1 Parameters used in the solution of equation sets (6.2) and (6.3).

Parameter	Value
L_{01}	0 nH
L_{C1}	908 nH
L_p	689 nH
$L_{th}\|_{z=0}$	144 nH
$\overline{L_{C1}}$	744 nH
C_1	4.5 μF
$V_{C1}\|_{t=0}$	28 kV
R_e	5.0 mΩ
R_p	0.5 mΩ
m_p	3.0 mg
$z(0)$	0.5 mm
$v_z(0)$	0 m/s

To further simplify the comparison of the two sets of equations, it was assumed that the ambient background gas density was zero, meaning all the mass m_p is in the plasma starting at time $t = 0$. The resulting computed solutions for these conditions and the data in Fig. 3.6 are presented in Fig. 6.1. We shall note that for this example, the values of initial voltage, capacitance, and plasma mass were selected without much effort expended on optimization. The results should therefore not be considered indicative of the general performance of inductive pulsed plasma thrusters.

While the general form of the solutions to equation sets (6.2) and (6.3) are similar, we observe that there are differences between the two sets, including a roughly 13% variance between the two calculated terminal velocity values. Most of this difference appears early in the acceleration process. While the inductance functions in Fig. 3.6 appear close to each other, we observe that the mutual inductance $\overline{M_{1p}}$ starts out higher than M_{1p} before both collapse to approximately the same curve. This difference is completely due to the manner in which the inductance functions were generated using the magnetic field solver. The solutions that comprise the inductance function from which $\overline{M_{1p}}$ is derived were obtained assuming that no external magnetic field could penetrate the accelerating annular disk. This is equivalent to stating that the current in the disk was flowing only on the surface of the disk facing the coil windings. The inductance calculations for M_{1p} were performed assuming that the current in the annular disk was uniformly distributed in the axial direction and varied as $1/r$ in the radial direction, leading to an inherently-lower mutual inductance in comparison to the concentrated surface current distribution. Collapsing the current density into a thinner sheet approaching a delta-function thickness pushes the mutual inductance M_{1p} towards $\overline{M_{1p}}$. This leads us to conclude that equation set (6.2) and equation set (6.3) yield roughly equivalent results, with any major differences in the solutions owing to variations in the assumed current density distribution used to calculate the inductance.

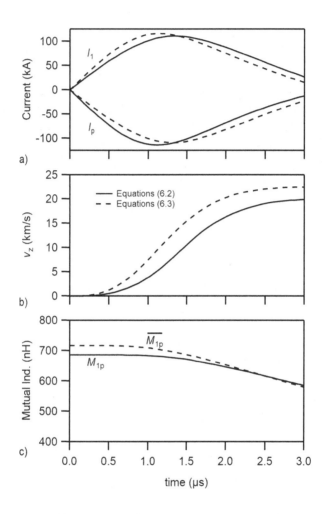

Figure 6.1 Computed solutions comparing results using the general equation set (6.2) [solid lines] and the simplified equation set (6.3) [dashed lines]. a) Currents in the external circuit (I_1) and the accelerated load (I_p), b) velocities of the accelerated load, and c) time evolution of the mutual inductance in each solution set.

6.2.2 Coilguns

The governing equation set (6.1) for a solid non-collapsing projectile, or macron, has been applied by Polzin *et al.* [6]

to the problem of modeling a coilgun. In the coilgun, illustrated schematically in Fig. 1.5, an annular metallic projectile is accelerated by a series of multiple coils aligned along the centerline. The projectile has a smaller diameter than the coils, and as it is accelerated in the axial direction it passes each coil in sequence. The switching of these coils is timed to initiate current flow as the projectile is passing through the coil.

Solutions to this problem were generated for a set where the mutual inductance terms between coils was handled in a self-consistent manner as described in this document and the reference. Presented in Fig. 6.2 are data from what Ref. [6] calls "case 3" for a single-turn projectile. Case 3 is defined as a solution where the capacitance for each external circuit, or stage, was varied so the projectile transit time between coils equaled the nearest active coil's first half-cycle. However, the

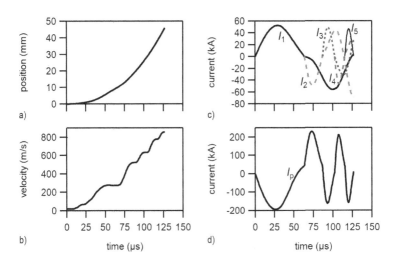

Figure 6.2 Computed solutions from a coilgun simulation (after Ref. [6]) showing solid conducting projectile a) axial position, b) axial velocity, c) current in each of the external coils, and d) current in the projectile.

discharge energy per coil stage was held constant throughout the problem, resulting in much higher initial capacitor voltages for later stages.

We observe clearly in these data the effect on the coil current I_p as each sequential coil is activated. These data also demonstrate the mutual inductance coupling between coils, with the current in the coils deviating from their trajectory as each nearby coil is activated. As an interesting aside, in systems such as the one illustrated where there are multiple independently-operating coils, the net force on the projectile acts to accelerate it in the axial direction. However, we observe that the currents in some of the coils are flowing in the same direction as that in the projectile, resulting in the exertion of a small but finite negative force on the projectile that slightly reduces the overall net axial force.

6.2.3   Axial Inductively-Coupled Planar Accelerators

There have been several studies using the lumped-element circuit model to capture the response of an axial inductively-coupled planar plasma accelerator where a single coil acts upon a flat annular plasma disk. For examples of this work, see Refs. [1, 2, 11, 12, 26] and the references therein. Almost all of these studies have been performed using equation set (6.3), which we previously showed was less accurate than the general formulation presented in equation set (6.2). Fixing those deficiencies in the literature data is beyond the scope of the present monograph, and will be left to future researchers in the field. However, even with a less accurate model, we can still show how various trends in data were captured by that modeling, which has been incredibly insightful in building a greater understanding of these accelerators.

6.2.3.1   Nondimensional Analysis

Polzin et al. [18, 27] have performed analyses of axial accelerators through nondimensionalization of equation set (6.3). The outcome was a set of dimensionless scaling parameters that govern the overall response of the coupled system. The parameters that emerged from that work were

$$L^* = \frac{L_0}{L_C}, \tag{6.6a}$$

$$\psi_1 = R_e \sqrt{\frac{C}{L_0}}, \tag{6.6b}$$

$$\psi_2 = R_p \sqrt{\frac{C}{L_0}}, \tag{6.6c}$$

$$\alpha = \frac{C^2 V_0^2 L_C}{2 m_p z_0^2}. \tag{6.6d}$$

The subscripts 1 have been dropped here for simplicity since we are discussing a system with only one external coil and one accelerated plasma sheet.

The first term in Eq. (6.6) is the inverse of the ratio of the potential change in external circuit inductance during an acceleration pulse over the stray inductance in the external circuit. This ratio is a means to quantify the electrical efficiency of the accelerator, corresponding roughly to the fraction of energy that can be directed into electromagnetic acceleration.

The terms ψ_1 and ψ_2 control the oscillatory nature of the current waveform in the external circuit and the plasma sheet, respectively. The discharge current waveforms in RLC circuits generally take the form of damped sinusoidal oscillations, with values below unity representing underdamped systems and values above unity representing overdamped systems.

The dynamic impedance parameter α (which is different from the parameter α found earlier in the equations governing conservation of mass, momentum, and energy) is probably the most important of the dimensionless parameters. It can be recast as

$$\alpha = \frac{C^2 V_0^2 L_C}{2 m_p z_0^2} = \frac{1}{8\pi^2} \frac{C V_0^2 / 2}{m_p v_z^2 / 2} L^* \left(\frac{2\pi \sqrt{L_0 C}}{L_0 / \dot{L}} \right)^2 \tag{6.7}$$

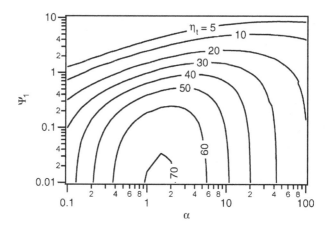

Figure 6.3 Contours of calculated axial inductively-coupled planar accelerator efficiencies (after Ref. [18]) for a slug mass loading as a function of α and ψ_1.

where V_0 is the initial charge voltage on the capacitor and $\dot{L}$ is the dynamic impedance. The parameter α has been shown to represent a ratio of the electromagnetic decoupling timescale and the period of the external circuit.

Presented in Fig. 6.3 are contours of constant efficiency (ratio of kinetic energy imparted to the accelerated plasma to initial energy stored in the capacitor) as a function of α and ψ_1. We observe that for a given value of ψ_1 there exists a maximum efficiency. This maximum corresponds to a situation where the oscillatory timescale of the external circuit is matched to the electromagnetic coupling timescale, permitting the greatest transfer of stored electrical energy into the kinetic energy of the gas. We also notice that for a fixed α, the efficiency increases with decreasing ψ_1. This emphasizes the importance of highly ringing external circuits with fast current rise rates in inductively-coupled pulsed accelerators.

6.2.3.2 Comparison with Data

The axial inductively-coupled planar accelerator is the one class of device for which sufficient data exist to perform real comparisons with modeling techniques. The largest data set exists on the Pulsed Inductive Thruster (PIT) MK V and MK Va [11]. The model of equation set (6.3) was applied to these data in Refs. [2, 11] and exhibited good quantitative agreement with the measured plasma exhaust velocity and efficiency. To demonstrate this, experimentally measured efficiencies, given as a function of α for the experimental condition, are presented in Fig. 6.4 and compared to solutions to the equation set generated for the design parameters of both the PIT MK V and MK Va thrusters. At the operational level, while the measured and simulated values of the discharge current given in Fig. 6.5 do not exactly align owing to a lack of ionization physics in the model and differences between the actual current density distribution and that assumed when calculating the inductance function, the current waveforms in the first half-cycle do show reasonably good agreement.

It is interesting to note that the MK Va was the only accelerator in the PIT series of devices to reach the maximum efficiency. One can see in the plots in Figs. 6.3 and 6.4 why the MK Va was able to achieve this level of performance and

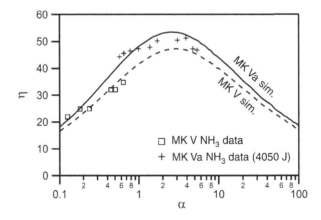

Figure 6.4 Comparisons of PIT MK V and MK Va performance data with efficiencies computed using equation set (6.3), displayed as a function of α (after Ref. [2]).

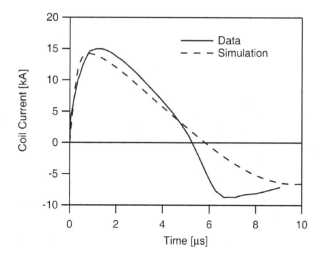

Figure 6.5 Comparisons of experimental (solid line, from [28]) and numerically-calculated current waveforms for a pulse in one (of nine) sets of coils in the PIT MK Va (from [29]).

why thrusters like the PIT MkV were unable to reach this maximized efficiency: the value of α was such that the ringing electrical source and the accelerating plasma load were well matched in the former but not in the latter.

6.2.4   Conical Theta-Pinch Plasma Accelerators

Martin [4] modeled a two-dimensional conical theta pinch plasma accelerator where the angle of the theta pinch coil θ_c was varied from 10 to 90 degrees (0° being a straight theta-pinch coil and 90° being a planar coil). The plasma was the same length as the accelerator coil and was held at the same angle as the coil for the duration of a simulation run. In this work a single external coil-plasma load system was modeled using equation set (6.4) (with some minor modifications). A slug loading of the plasma was assumed, eliminating any complications associated with interactions between the moving plasma and the ambient gas.

Computed results for the case of a 60° cone angle are shown in Fig. 6.6. The waveform for the external circuit

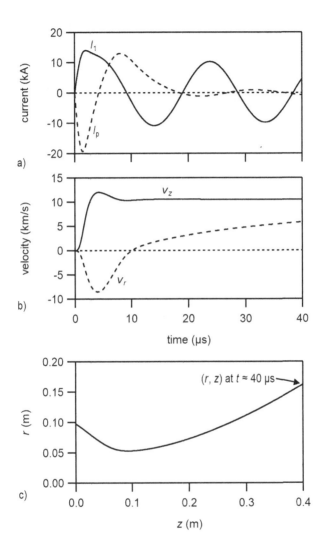

Figure 6.6 Computed solutions for a conical theta-pinch plasma accelerator with a coil angle θ_c of 60°, showing time-histories of: a) the external circuit and plasma load currents, b) the radial and axial velocities, and c) a trajectory of the plasma in two-dimensional r-z space (after Ref. [4]; © IOP Publishing. Reproduced with permission. All rights reserved.)

current I_1 is observed to change in form after just 1-2 μs. This elongation of the driving current in time is indicative of the performance of inductively-driven electromagnetic work. The axial velocity v_z of the plasma peaks at about 4 μs and after slightly decreasing it remains constant for the rest of the pulse. The radial plasma velocity v_r is interesting as it initially decreases, but the effect of its self inductance causes it to begin expanding at about 4 μs. This leads to a computed path for the plasma in the r-z plane that first compresses and then expands, looking very much like the notional conception of this path previously shown in Fig. 4.1b.

CHAPTER 7

Concluding Remarks

C IRCUIT modeling techniques can be a very powerful set of tools in the study of inductively-coupled pulsed accelerators. While they are not as complex as finite element fluid solvers or particle-in-cell methods, lumped-element circuit modeling has proven to be quite insightful in the study of these devices.

The assumed geometry of the problem and how the conducting body is constrained to move in the r-z plane are very important in deciding which terms must be included in the circuit equations and the equations of motion for the system. There are many different types of inductively-coupled pulsed accelerators, but through some tailoring to account for the variations between accelerator types it is possible to model all of these systems using lumped-element circuit modeling techniques.

It was shown how different assumptions resulted in slight but finite differences in terms like the computed mutual inductance between a stationary external circuit and a moving conductor, and how those small differences could compound into larger deviations in the course of numerically solving the coupled set of equations.

Solutions from the literature were presented for various inductively-coupled pulsed accelerator geometries and configurations. There exist a limited number of studies that have employed this simulation and modeling method. For those

DOI: 10.1201/9780429351976-7

cases where experimental data exist, modeling results compare quite favorably. For axially-accelerating plasma sheets, the model (in non-dimensional form) not only explains the observation of a maximum in measured accelerator efficiency, but also why that maximum was only observed in PIT MK Va data while data from other PIT designs did not achieve a maximum.

In comparing solutions generated using the lumped-element circuit model to experimental data, one issue that arises is incompleteness of the experiment description. If the description of a study is missing critical dimensions or values of certain parameters, it can be nearly impossible to model the system accurately. Moreover, even if the description is complete, a comparison with modeling results is only possible if experimentally measured data that is comparable to the outputs of the model exist. As described in this monograph these data exist for the PIT series of devices, but for many of the other inductive accelerator variants this is not the case.

Lumped-element circuit modeling has already proven to be a useful and insightful tool in understanding inductively-coupled pulsed accelerators, and as more data become available and more researchers apply these modeling methods to a greater number of complex problems, the usefulness of the technique can only increase.

Bibliography

[1] K Polzin, A Martin, J Little, C Promislow, B Jorns, and J Woods. State-of-the-art and advancement paths for inductive pulsed plasma thrusters. *Aerospace*, 7(8):105, 2020.

[2] K A Polzin. Comprehensive review of planar pulsed inductive plasma thruster research and technology. *Journal of Propulsion and Power*, 27(3):513–531, 2011.

[3] A K Hallock. *Effect of Inductive Coil Geometry on the Operating Characteristics of a Pulsed Inductive Plasma Accelerator*. PhD thesis, Princeton University, 2012. Thesis number 3252-T.

[4] A K Martin. Performance scaling of inductive pulsed plasma thrusters with coil angle and pulse rate. *Journal of Physics D: Applied Physics*, 49(2):025201, 2016.

[5] K A Polzin, J E Adwar, and A K Hallock. Optimization of electrodynamic energy transfer in coilguns with multiple, uncoupled stages. *IEEE Transactions on Magnetics*, 49(4):1453–1460, 2013.

[6] K A Polzin, A B Cipriano, A K Martin, and C Y Liu. Coilgun acceleration model containing multiple interacting coils. In *AIAA Scitech 2019 Forum*, 2019. AIAA paper 2019-1247.

[7] C S Hill. *Translation Studies on an Annular Field Reversed Configuration Device for Space Propulsion*. PhD thesis, Michigan Technological University, 2012.

[8] J M Woods, C L Sercel, T M Gill, E Viges, and B A Jorns. Data-driven approach to modeling and development of a 30 kW field-reversed configuration thruster. In *36th International Electric Propulsion Conference*, 2019. IEPC paper 2019-717.

[9] J M Woods, C L Sercel, T M Gill, and B A Jorns. Equivalent circuit model for a rotating magnetic field thruster. In *AIAA Propulsion and Energy 2021 Forum*, 2021. AIAA paper 2021-3400.

[10] A K Martin, A Dominguez, R H Eskridge, K A Polzin, D P Riley, and K A Perdue. Design and testing of a small inductive pulsed plasma thruster. In *34th International Electric Propulsion Conference*, 2015. IEPC paper 2015-50.

[11] C L Dailey and R H Lovberg. The PIT MkV Pulsed Inductive Thruster. Technical Report NASA CR 191155, TRW Space & Technology Group, Redondo Beach, CA, 1993.

[12] R H Lovberg and C L Dailey. A PIT primer. Technical Report TR 005, RLD Associates, Lebanon, PA, 1994.

[13] K A Polzin, A K Martin, R H Eskridge, A C Kimberlin, B M Addona, A P Devineni, N R Dugal-Whitehead, and A K Hallock. Summary of the 2012 inductive pulsed plasma thruster development and testing program. Technical Report NASA/TP-2013-217488, NASA-Marshall Space Flight Center, Huntsville, AL, 2013.

[14] B M Novac, I R Smith, M C Enache, and P Senior. Studies of a very high efficiency electromagnetic launcher. *Journal of Physics D: Applied Physics*, 35(12):1447–1457, 2002.

[15] L Shoubao, R Jiangjun, P Ying, Z Yujiao, and Z Yadong. Improvement of current filament method and its applica-

tion in performance analysis of induction coil gun. *IEEE Transactions on Plasma Science*, 39(1):382–389, 2011.

[16] A Shimazu and J Slough. Simulation of electromagnetic inductive drive for liner implosion fusion thruster system using filamentary approaches. In *35th International Electric Propulsion Conference*, 2017. IEPC paper 2017-506.

[17] K A Polzin, K Sankaran, A G Ritchie, and J P Reneau. Inductive pulsed plasma thruster model with time-evolution of energy and state properties. *Journal of Physics D: Applied Physics*, 46(475201), 2013.

[18] K A Polzin and E Y Choueiri. Performance optimization criteria for pulsed inductive plasma acceleration. *IEEE Transactions on Plasma Science*, 34(3):945–953, 2006.

[19] D J Griffiths. *Introduction to Electrodynamics, 4th Edition*. Cambridge University Press, Cambridge, UK, 2017.

[20] S. Chandrasekhar. *Hydrodynamic and Hydromagnetic Stability*. Dover, New York, 1981.

[21] J Slough, A Pancotti, and D Kirtley. Analysis of inductively driven liners for the generation of megagauss magnetic fields. In *14th International Conference on Megagauss Magnetic Field Generation and Related Topics (MEGAGAUSS)*, 2012.

[22] E C Cnare. Magnetic flux compression by magnetically imploded metallic foils. *Journal of Applied Physics*, 37(10):3812–3816, 1966.

[23] D Kirtley, J Slough, J Schonig, and A Ketsdever. Pulsed inductive macron propulsion. In *57th Joint Army-Navy-NASA-Air Force (JANNAF) Spacecraft Propulsion Subcommittee Meeting*, 2010.

[24] F D Murnaghan. The compressibility of media under extreme pressures. *Proceedings of the National Academy of Sciences*, 30(9):244–247, 1944.

[25] A C Robinson. The Mie-Grüneisen power equation of state. Technical Report SAND2019-6025, Sandia National Laboratory, 2019.

[26] R H Lovberg and C L Dailey. Large inductive thruster performance measurement. *AIAA Journal*, 20(7):971–977, 1982.

[27] K A Polzin. Pulsed inductive plasma acceleration: Optimization. In J Shohet, editor, *Encyclopedia of Plasma Technology*, pages 1191–1200. CRC Press, Boca Raton, FL, 2016.

[28] I Hrbud, M LaPointe, R Vondra, C L Dailey, and R Lovberg. Status of pulsed inductive thruster research. *AIP Conference Proceedings*, 608:627–632, 2002.

[29] K A Polzin. *Faraday Accelerator with Radio-Frequency Assisted Discharge (FARAD)*. PhD thesis, Princeton University, 2006. Thesis number 3147-T.

Index

Printed in the United States
by Baker & Taylor Publisher Services